ISBN 978-1-330-06472-6
PIBN 10016820

This book is a reproduction of an important historical work. Forgotten Books uses
state-of-the-art technology to digitally reconstruct the work, preserving the original format
whilst repairing imperfections present in the aged copy. In rare cases, an imperfection in
the original, such as a blemish or missing page, may be replicated in our edition. We do,
however, repair the vast majority of imperfections successfully; any imperfections that
remain are intentionally left to preserve the state of such historical works.

1 MONTH OF
FREE
READING

at

www.ForgottenBooks.com

By purchasing this book you are eligible for one month membership to ForgottenBooks.com, giving you unlimited access to our entire collection of over 1,000,000 titles via our web site and mobile apps.

To claim your free month visit:

www.forgottenbooks.com/free16820

English
Français
Deutsche
Italiano
Español
Português

www.forgottenbooks.com

Mythology Photography **Fiction**
Fishing Christianity **Art** Cooking
Essays Buddhism Freemasonry
Medicine **Biology** Music **Ancient**
Egypt Evolution Carpentry Physics
Dance Geology **Mathematics** Fitness
Shakespeare **Folklore** Yoga Marketing
Confidence Immortality Biographies
Poetry **Psychology** Witchcraft
Electronics Chemistry History **Law**
Accounting **Philosophy** Anthropology
Alchemy Drama Quantum Mechanics
Atheism Sexual Health **Ancient History**
Entrepreneurship Languages Sport
Paleontology Needlework Islam
Metaphysics Investment Archaeology
Parenting Statistics Criminology
Motivational

The AMATEUR MECHANIC

BY

A. FREDERICK COLLINS

AUTHOR OF "KEEPING UP WITH YOUR MOTOR CAR," "HOW TO FLY,"
"THE BOOK OF ELECTRICITY," ETC.

FULLY ILLUSTRATED

D. APPLETON AND COMPANY
NEW YORK   LONDON
1918

TO
MY NIECE AND NEPHEW
ETHEL AND EARL COLLINS

A WORD TO YOU

Don't do anything until you have read this book!

I might qualify the above statement by saying that if you are an amateur it will pay you to scan the following pages before you try to do mechanical things.

The idea I have tried to carry out is to parallel the case of the locomotive engineer. You know, of course, that he did not build the engine he drives but he knows every part of it, exactly how it works, how to run it to get the most power, or speed, or both, out of it with the highest fuel economy and, further, if he should have a breakdown on the road he knows just how to make whatever repairs are needed to go on with his run.

I have presupposed that you know how to use ordinary tools (though I have explained the mode of operation of a few that relate to the art of measuring) and I have not told how to make the various devices and machines described but what I have gone into is how things are constructed, how to make simple calculations to get the result you want, how the machine works, how to run it to get the most light, heat or power out of it at the least cost for fuel, upkeep and expenditure of labor, how to repair it when something happens, and, lastly, how to buy it.

A further purpose of this book is to tell about the kinds of materials that are used in building and the appliances that are employed in operating a home or a farm so that if you are a householder or a husbandman you can enjoy all the benefits of the electrical and mechanical arts known that make for the comfort, convenience, economy and safety of yourself and family and so make life worth living.

A. FREDERICK COLLINS.

600 Riverside Drive,
  New York City.

CONTENTS

CONTENTS

CONTENTS

LIST OF ILLUSTRATIONS

LIST OF ILLUSTRATIONS

LIST OF ILLUSTRATIONS

THE
AMATEUR MECHANIC

CHAPTER I

RULES AND TOOLS FOR MEASURING

All tools for measuring may be divided into two classes, and these are (1) *rules* and *instruments* for making actual measurements, and (2) *gauges* for testing and comparing.

A rule is simply a strip of wood, or metal, or other material, having a straight edge and whose surface is graduated into inches or *centimeters* [1] and fractions thereof. This graduated surface is called a *scale*, and sometimes the rule itself is spoken of as a scale.

A Carpenter's Boxwood Rule.—Carpenters' rules are not all made alike, for some are 1 foot-4 fold, some are 2 foot-2 fold, those in general use are 2 foot-4 fold, others are 3 foot-4 fold and, finally, there are 4 foot-4 fold rules.

But a regular carpenter's rule is taken to mean a 2 foot-4 fold boxwood rule, the scales being divided into eighths, tenths, twelfths and sixteenths of an

[1] A unit of lineal measurement used in the *Metric System.*

inch. To measure closely, turn the rule on its edge so that the graduated lines set against the board or whatever it is you are measuring. A rule of this kind is shown at A in Fig. 1.

The Triangular Boxwood Rule and Scale.—If you are making machine or architectural drawings you should by all means have one of these scales, for with it you can draw *to scale,* or get the actual dimensions from drawings that have been made to scale, both easily and quickly.

This rule, which is shown at B, has, as you will see, *three sides* and each side has *two surfaces,* making *six surfaces* in all. On one of these surfaces there is an ordinary twelve-inch scale graduated in inches and a different scale is graduated on each end of the other five surfaces, thus making eleven scales all told. These other ten scales are graduated to $\frac{3}{32}$, $\frac{3}{16}$, $\frac{1}{8}$, $\frac{1}{4}$, $\frac{3}{8}$, $\frac{1}{2}$, $\frac{3}{4}$, 1, $1\frac{1}{2}$ and 3 inches to the foot.

To Learn the Rule.—Lay it on the table with the twelve-inch scale away from you, just as though you were going to draw a line and so that it reads from 0 on the left to 12 on the right. Now turn the rule toward you until the next side is uppermost, and you will see that the upper left-hand scale reads to $\frac{3}{4}$ of an inch toward the right, and that the upper right-hand scale reads to $\frac{3}{8}$ of an inch toward the left.

The lower left-hand scale, you will observe, reads to 3 inches toward the right, and that the lower right-hand scale reads to $1\frac{1}{2}$ inches toward the left. You will also note that the left-hand upper and lower

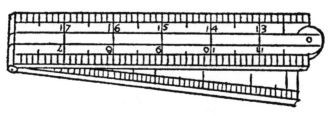

A   CARPENTER'S RULE

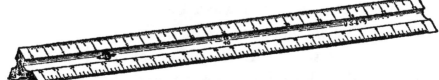

B   ARCHITECT'S SCALE

C   PATTERN MAKERS RULE

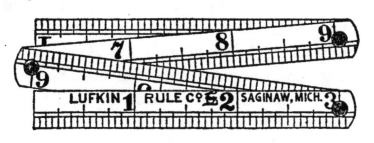

D   A FLEXIBLE RULE

Fig. 1.—Rules and Scales

scales are just twice as long as their respective right-hand scales.

Further at the beginning of each scale there is a space marked into either 12 or 24 smaller spaces. These spaces represent 1 foot, or 12 inches, and each space represents $\frac{1}{2}$ inch when there are 24 of them, or 1 inch when there are 12 of them.

How to Use the Rule.—Now suppose you are drawing the plans for a drawer and that you want your plan when it is done to be $\frac{3}{8}$ as large as the drawer will really be, that is, you want your plans drawn to a scale so that $\frac{3}{8}$ inch will equal 1 inch.

If the drawer is to be $7\frac{1}{2}$ inches wide and you drew it with an ordinary rule, you would have to multiply $\frac{3}{8}$ by $7\frac{1}{2}$ and the product will tell you that you must measure off $4\frac{5}{8}$ inches or forty-five $\frac{1}{16}$ inch spaces from the end. And you would do this two or three times over, because you would be the exception if you didn't lose count of them.

By using the triangular rule you do away with all this bother, for, if in the drawing you are making $\frac{3}{8}$ inch equals 1 inch and the width is $7\frac{1}{2}$ inches, you simply use the scale marked $\frac{3}{8}$ inch and mark off $7\frac{1}{2}$ spaces. Or if your drawing is to a scale of $\frac{3}{8}$ inch to 1 foot you start at the 0 end of the $\frac{3}{8}$ inch scale and measure off 7 big spaces to the left; next measure off 6 of the small spaces to the right of 0, since each of these represents 1 inch and since 6 inches equals $\frac{1}{2}$ a foot; and thus you have measured off $7\frac{1}{2}$ feet with the scale where $\frac{3}{8}$ equals 1 foot.

In the same way you can use any scale on the rule and make working drawings to any scale within its limitations and without any calculation whatever. The chief thing to remember is that each of these scales starts off with a space divided into 12 parts or 24 parts depending on the size of the scale and whether this space represents 1 foot and the smaller spaces ½ an inch or 1 inch, as the case may be. A rule of this kind can be bought [2] for as little as sixty cents.

A Pattern Maker's Shrinkage Rule.—When a casting is made the metal shrinks on cooling, and to allow for this shrinkage the pattern must be made a little larger than the casting is to be.

A shrinkage rule, see C, is graduated to allow for the shrinkage of the metal you are using. The spacing of the graduations is used to measure the patterns you are making, while the figures on the graduations show the actual size the castings will be.

The Use of Flexible Rules.—Rules made of cardboard, celluloid, thin steel and wood are useful for measuring curved surfaces.

Cardboard rules can be bought [3] for a couple of cents each; *celluloid rules* 6 inches long can be had for five cents each, and very thin spring-tempered rules for machinists [4] can be purchased in any lengths

---

[2] Triangular boxwood rules can be bought of the L. E. Knott Apparatus Co., Boston, Mass., and also of Keuffel and Esser Co., 127 Fulton St., New York.

[3] The L. E. Knott Apparatus Co. sells these.

[4] These rules are sold by Hammacher, Schlemmer and Co., Fourth Ave. and 13th St., New York.

from 1 inch up to 48 inches for fifteen cents for the shortest up to $7 for the longest.

Where measurements of doors, windows, boilers,

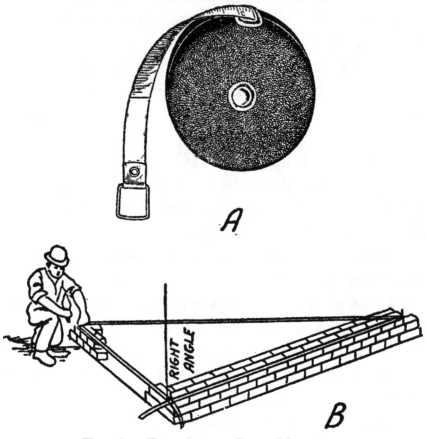

FIG. 2.—THE STEEL TAPE MEASURE

etc., are to be made a flexible *folding wood rule* will be found very convenient, while a flexible steel folding rule, as shown at D, is a good one to use for metal work.

About Tape Measures.—These elongated rules are used by every carpenter, mason, contractor, surveyor and engineer and you ought to have one in your kit of tools.

An ordinary tape measure, see A in Fig. 2, consists of a thin, flexible steel tape from $\frac{1}{4}$ to $\frac{5}{8}$ inch wide and from 25 to 100 feet long; it is graduated on one side into feet, inches and eighths and is fitted into a hard leather case. The tape can be reeled up by a handle which folds in flush with the side of the case.

The *Roe tape measure* has a right angle attachment which permits it to be used quickly and accurately for laying out right angles as shown at B. It is based on the well-known trigonometrical formula that a triangle whose sides measure 6, 8 and 10 feet makes a right angle. Hence, by using this tape measure you can get a perfect right angle without a surveying instrument or tools or help of any kind.

The Carpenter's Steel Square.—The ordinary *carpenter's square,* or *steel square,* or *framing square,* as it is variously called, is used not only as a rule, a straight edge and a try square in building construction but also for laying out octagons, or *8 squares,* as they are called, finding the square feet in boards, or *board measure* as it is termed, finding the lengths and cuts of braces and also of common, hip, valley and jack rafters for different pitches of roofs.

The ordinary steel square is formed of two parts though these are usually made of one piece of steel about $\frac{1}{8}$ of an inch thick and which set at right angles to each other as shown at A in Fig. 3. The long

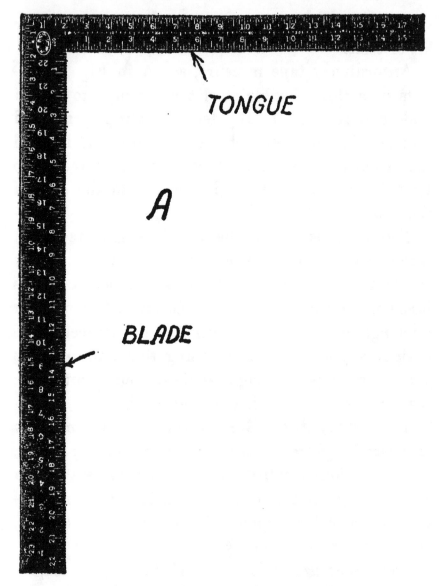

TONGUE

A

BLADE

FIG. 3A.—THE CARPENTER'S STEEL SQUARE

piece of the square is called the *blade* and is about 2 inches wide and 24 inches long; the short piece is called the *tongue* and this is about 1½ inches wide and 16 inches long.

The side of the square with the maker's name stamped on it is called its *face* and the other and opposite side is called its *back*. It is usually divided into $\frac{1}{100}$, $\frac{1}{32}$, $\frac{1}{16}$, $\frac{1}{12}$, $\frac{1}{10}$ and $\frac{1}{8}$ inch scale divisions.

Laying Out an Octagon or 8-Square.—Along the middle of the tongue of the square you will find

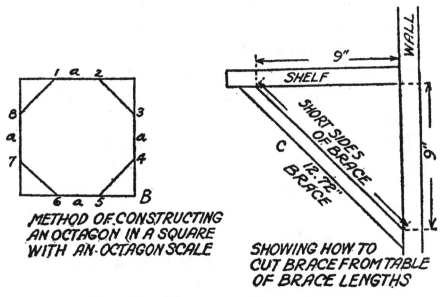

METHOD OF CONSTRUCTING AN OCTAGON IN A SQUARE WITH AN OCTAGON SCALE

SHOWING HOW TO CUT BRACE FROM TABLE OF BRACE LENGTHS

FIG. 3B.—USING THE STEEL SQUARE

a scale for drawing octagons, or 8-squares, as shown at B in Fig. 3.

To draw an octagon you must first draw a square which is just large enough to contain the sized octagon you are going to make. Having drawn your square, *bisect* it, that is, find the middle of each side of it with your dividers, as at (a) which is the middle point.

Next take your dividers and, using the 8-square

scale of the square, set them to as many spaces of the scale as there are inches in any one side of the square you have drawn. Lay this distance off on both sides of each middle point, as shown at B 1, 2, 3, 4, 5, 6, 7 and 8. Then connect these points with lines starting at 1 and drawing to 2, from 2 to 3, and so on until the octagon is complete.

You will observe that it is the 8-square scale on the square and not the square itself that is the important part of laying out octagons and that a scale of this kind marked on a rule would serve the purpose just as well.

The Brace Measure Table.—Along the center of the back of the tongue of your square you will find a table of numbers and you will see that there are two numbers, one above the other, which are equal and one number to the left of them.

The purpose of this table is to make it possible for a carpenter to instantly determine the length of a brace when its ends are to be fixed at equal distances from the intersecting post, beam, shelf, wall or any other like construction.

The table is used like this: Suppose that you have a shelf you want to fix to the wall with a pair of braces, and that you want to have each end of each brace 9 inches from the point where the wall and the shelf intersect each other. Look at the table and you will see that after the set of number 9 the number 12.72 is just to the left of it.

This number—12.72—is the length in inches, then,

that you must make the short side of the brace, so cut a piece of wood a little longer than 12.72 inches —say 15 inches—if the brace is to be made 12.72 inches on one side. Now put it in your miter box and cut off each end at an angle of 45 degrees, when it will just fit into the corner with each end 9 inches away from the intersection of the shelf and wall, as shown at C.

This table is based on the same trigonometrical relations between the lengths of the sides of a right triangle as that described under the caption of *Tape Measures*.

The Essex Board Measure Table.—The term *board measure* means the number of square feet in a board 1 inch thick. A board 2 inches thick will have twice as many *board feet* in it as a board 1 inch thick, and so on.

Of course a board 12 inches wide will have as many feet in it as it is feet long and you don't have to do any figuring to know the answer. But if the board is more or less than 12 inches wide you will have to make a small calculation to find the board feet in it. If, for instance, the board is 8 inches wide and 10 feet long, to figure out the board feet you will have to find the number of square inches in it first and then divide the product by 144.

But if you use the Essex board measure table on the square you can instantly find the number of board feet in a board without any calculation. The starting point in this table is always the figure 12. If, now, you want to find the board feet in a board 8 inches wide and 10 feet long, simply follow the graduated line on the left of the table down to the figure 10, then follow the cross line toward the *left* to 8, and you will find that the number under 8 is 6; you will also see that 6 is to the left of the cross line and 8 is to the right, which means that there is 6 feet and $\frac{8}{12}$ inches, board measure, in the board.

But if the board is wider than 12 inches, then you follow the cross line toward the *right* to the number representing the length of the board you want to measure. If the board is 2 inches, multiply the result you get by 2, which will give you the board measure for that thickness.

The Rafter Framing Table.—On the back of a good steel square you will find a table of numbers marked between the scales of inches on the tongue.

With this table you can find the *lengths* for rafters of known *rise* and *run* for a given *pitch*. The *rise* of a rafter is the vertical height from its ridge end to a horizontal line on a level with its foot.

The *run* of a rafter is the reach in length from the outside edge of its foot to a point exactly under its ridge end on a horizontal line level with its foot. The *pitch* of a rafter is the *ratio* of the rise to twice the run, which is usually equal to the width of the building.

Now, if you will look at the left of the table you will see a series of figures, thus:

PITCH

12—4	1/6
12—6	1/4
12—8	1/3
12—10	5/12
12—12	1/2
12—15	5/8
12—18	3/4

These numbers are the common *pitches* at which rafters are set. Now if your rafter is to be set at such a slant that for every horizontal foot of run, or 12 inches, it rises 4 inches vertically, then the pitch, or slant, of the rafter is $\frac{1}{6}$.

Let's suppose you have a rafter with a known pitch of $\frac{1}{6}$ and a run of 4 feet and you want to find the length. Follow the horizontal line on the table on your square from $\frac{1}{6}$ until you come under the 4 on the inches scale at the top of the table, and the number you find under 4 will be the length of the rafter needed, thus:

	3		4	
Pitch				
12—6	1/64	2	7	

Hence, for a rafter having a pitch of $\frac{1}{6}$ and a run of 4 feet, the required length will be 4 feet, 2 inches and $\frac{7}{12}$ of an inch. The first number 4 represents the feet, the second inches and the third twelfths of an inch. In the same way you will see that a rafter

• 13

having a $\frac{1}{3}$ pitch and a run of 20 feet will be 24' 0" $\frac{6}{12}$", or 24 feet and $\frac{1}{2}$ inch.

The Vernier.—It is easy to measure small fractions of an inch with the *vernier* which cannot be measured at all with an ordinary rule.

The vernier, as shown at A, in Fig. 4, consists of

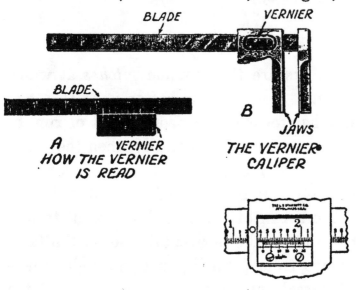

A
HOW THE VERNIER IS READ

B
THE VERNIER CALIPER

C
MICROMETER VERNIER READING TO 10,000 THS OF AN INCH

Fig. 4.—The Vernier

a short rule the scale of which slides against the scale of an ordinary rule. Because the scale divisions of the vernier and rules are of different widths, it is possible to read very small parts of the spaces with a good deal of accuracy. It is called a vernier after Pierre Vernier, the French mathematician, who invented it.

The principle on which the vernier works is this: First the scale of the ordinary rule is divided, let's say, into 10ths of an inch, and that the vernier is 1 inch long and is divided into 9ths of an inch—that is, it has one less scale division or space to the inch than the rule it slides against.

Now when the end lines of both the scales of the vernier and the rule meet, that is, when they are in a line with each other, the 10th line on the vernier will exactly coincide with the 9th line on the rule.

If, now, you slide the vernier toward the right so that the first lines on the vernier and rule meet, the vernier will have moved exactly $\frac{1}{10}$ of a scale division or space, which is $\frac{1}{100}$ of an inch, for this is the difference between the two scales. By sliding the vernier over until the second lines meet, it will have moved $\frac{2}{10}$ of a scale division, or $\frac{2}{100}$ of an inch, or $\frac{1}{50}$ of an inch, and so on. Verniers are put on and used with various measuring tools, such as calipers, protractors, etc.

The Vernier Caliper.—The *vernier caliper* shown at B is made to take inside as well as outside measurements. It is graduated on the front to read, by means of the vernier, to 1000ths of an inch and on the back to 64ths of an inch.

How to Read a Vernier Caliper.—There are three chief makes of vernier calipers, and these are (1) the *Brown and Sharp,* (2) the *Starrett* and (3) the *Columbia Pattern.*

On either of the first two makes of calipers the

scale of the tool is graduated to 40ths, that is, in .025 ($\frac{1}{40}$) of an inch, and every fourth division, which is $\frac{25}{1000}$ of an inch, is numbered.

On the vernier plate there is a space divided into 25 parts, and these are numbered 0, 5, 10, 15, 20 and 25. These 25 divisions on the vernier take up exactly the same space as the 24 divisions on the scale of the rule. This makes the difference between the width of one of the 25 spaces on the vernier and one of the 24 spaces on the scale $\frac{1}{25}$ of $\frac{1}{40}$ or $\frac{1}{1000}$ of an inch.

If now the vernier is set so that the 0 line on the vernier coincides with the 0 line on the rule, the next two lines will be $\frac{1}{1000}$ of an inch apart, the next two lines will be $\frac{2}{1000}$ of an inch apart, and so on.

To read the caliper after having made a measurement, see how many inches, $\frac{1}{10}$ (or .100) and $\frac{1}{40}$ (or .025), the 0 mark on the vernier is from the 0 mark on the rule, and then count the number of divisions on the vernier from 0 to a line which exactly coincides with a line on the scale.

In the picture shown at C the vernier has been moved to the right 1 $\frac{4}{10}$ and $\frac{1}{40}$ inches, or 1.425 inches, as the 11th line on the vernier coincides with a line on the rule. $\frac{11}{1000}$ of an inch must in consequence be added to the reading on the scale of the rule and the total reading is therefore 1.436 inches, which is the distance the jaws of the caliper have been opened.

The Micrometer Caliper.—The *Micrometer caliper*, or just *micrometer* for short, is a little tool which will measure very accurately from 0 to 1 inch in thousandths or even ten-thousandths of an inch.

A micrometer is formed of (1) a *frame* to which is fixed (2) the *anvil* and through which (3) the *spindle* passes; the spindle is fastened to (4) the *thimble* and these turn in (5) the *sleeve,* as shown at **A** in Fig. 5.

How to Read a Micrometer.—To measure the thickness of a sheet of paper or anything else, put it between the anvil and the end of the spindle and hold the frame with your left hand. Now turn the thimble with your right hand and since the spindle is fixed to the thimble it of course turns with it. This makes it move through the nut in the frame and toward or away from the anvil.

The distance between the opposed surfaces of the anvil and the spindle is shown by the lines and figures on the sleeve and the thimble, and these tell the thickness of the thing you have measured.

The *pitch* of the screw threads on the inside part of the spindle which screws through the nut, is 40 to the inch; one complete turn of the spindle, therefore, moves it up or down $\frac{1}{40}$, or $\frac{25}{1000}$, of an inch. The sleeve is marked with 40 lines to the inch and these correspond to the number of threads on the spindle.

When the end of the spindle rests on the anvil the graduated edge of the thimble is exactly even with the line marked 0 on the sleeve and the 0 line on the thimble tallies with the horizontal line on the sleeve.

Now if you will open the micrometer by giving the thimble one full turn, or until the 0 line on the thimble again coincides with the horizontal line on the sleeve, the distance between the anvil and the sleeve is then $\frac{1}{40}$ of an inch, or .025 of an inch, and the graduated edge of the thimble will coincide with the second vertical line on the sleeve.

Each vertical line on the sleeve indicates a distance of $\frac{1}{40}$ of an inch; every fourth line is made

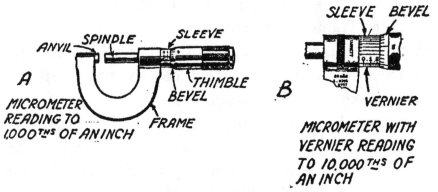

FIG. 5.—THE MICROMETER

longer than the others and is numbered 0, 1, 2, 3, etc., and each line so numbered indicates a distance of four times $\frac{1}{40}$ of an inch or $\frac{1}{10}$.

The graduated edge of the thimble is marked in 25 divisions with every fifth line numbered from 0 to 25. When you turn the thimble from one of these marks to the next, you move the spindle up or down $\frac{1}{25}$ of $\frac{25}{1000}$ or the $\frac{1}{1000}$ part of an inch. By turning it two divisions it shows two $\frac{1}{1000}$, etc., while 25 divisions shows one complete turn or .025 of an inch, or $\frac{1}{40}$ of an inch.

All you have to do to read the micrometer, then, is to multiply the number of vertical divisions which you can see on the sleeve by 25 and all the number of divisions on the graduated edge of the thimble from the 0 line to the line which tallies with the horizontal line on the sleeve; multiply this number by 25 and add the number of divisions shown on the

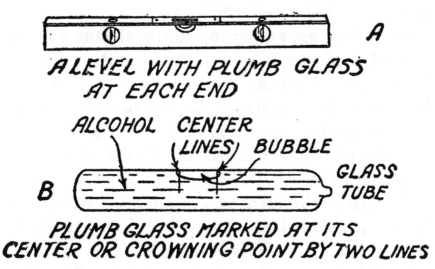

FIG. 6.—A LEVEL AND ITS PLUMB GLASS

bevel of the thimble. In the cut shown at A the micrometer is open $7 \times 25 = 175 + 3 = 178$ or $\frac{178}{1000}$ of an inch.

A Micrometer Reading to Ten-Thousandths.—A vernier is used on a micrometer, see B, in order to read it to $\frac{1}{10,000}$ of an inch. To read a $\frac{1}{10,000}$ micrometer, first find the thousandths of an inch as described above, and then note the line on the thimble. If it is the second line, marked 1, add $\frac{1}{10,000}$; if it is the third line, marked 2, add $\frac{2}{10,000}$, etc.

Gauges for Testing and Comparing.—Gauges for every purpose to facilitate or to make more accurate the work of the mechanic can be bought at almost any hardware store. If you cannot get what you want, write *Hammacher, Schlemmer & Co.,* Fourth Avenue and 13th Street, New York, and they will most likely be able to supply you with the tool you need.

One of the most common and useful gauges is the *carpenter's or mason's level,* shown at A in Fig. 6. When you are putting in a foundation for either a building or for machinery, the first thing to do is to find whether the top of it is level. This is done with a level; and to ascertain whether the side of the wall is *plumb,* an *upright level,* or *plumb,* must be used.

A spirit level consists of a sealed glass tube nearly filled with alcohol and having a bubble floating in it, as shown at B. This *plumb glass,* as it is called, is set in a *stock,* or length of wood, when the whole device is called a *level.* When the level is laid on a level surface the bubble will be in the middle of the glass, but if the surface is not level the bubble will flow to one end or the other to indicate it.

Levels are usually made with two plumb glasses, one in the upper edge and one in the top of and at right angles to it, so that it can be used to find if the side of a wall, as well as the top of it, is level. A few of the more common gauges used by machinists are shown in Fig. 7.

The Protractor.—To find any angle or to plot one

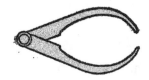

A-OUTSIDE CALIPERS

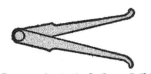

B-INSIDE CALIPERS

C SCREW THREAD GAUGE

D- WIRE GAUGE

E- DEPTH GAUGE

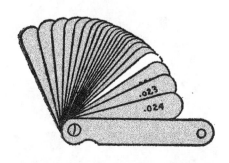

F-THICKNESS GAUGE

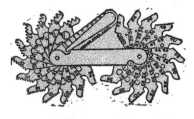

G-THREAD, SCREW AND TWIST DRILL GAUGE

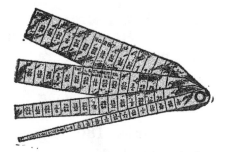

H- TAPER GAUGE

FIG. 7.—A FEW OTHER USEFUL GAUGES

from 0 to 360 degrees, a *protractor* is used. This instrument is usually made in the shape of a semi-circle and, as there are 360 degrees in a circle, there

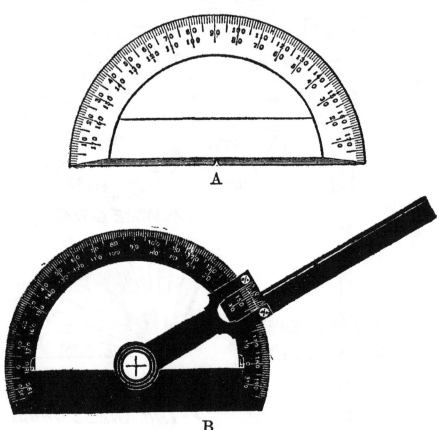

A

B

FIG. 8.—THE PROTRACTOR FOR FINDING ANGLES AND MEASURING THEM IN DEGREES

A.—A Brass Protector graduated in single degrees

B.—A German Silver Protector graduated in ½ degrees with vernier arm reading to 1 minute

are, of course, 180 degrees in a semicircular protractor. Each degree can be further divided into 60 *minutes* and each minute into 60 *seconds*, like the hour in our time system.

A brass protractor 3¾ inches in diameter can be bought[5] for as little as 25 cents. One of this kind is shown at A in Fig. 8. For all ordinary work scale divisions of 1 degree, or ½ degree, will be found close enough; but where readings to minutes are needed a *vernier protractor,* as shown at B, must be used.

To use an ordinary protractor, place it on a sheet

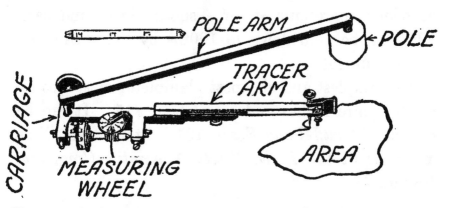

FIG. 9.—A CHEAP PLANIMETER FOR MEASURING THE AREA OF ANY PLANE SURFACE

of paper, lay a rule on top of it and keep its edge exactly over the nick in the middle of its straight edge. Then move the edge of the rule until it is on the line of the degree you want to mark off. Draw a line and you will have the angle you want.

The Planimeter.—This instrument gets its name from *planus,* which is Latin for *flat,* and *meter,* which comes from the Greek *metron,* meaning to measure. It is shown in Fig. 9.

It is so constructed that by a simple mechanical

[5] L. E. Knott Apparatus Co., Boston, Mass.

operation the area of any flat figure, however irregular the boundary line of it may be and drawn to whatever scale, such as a plot of ground, plans, indicator diagrams, etc., can be easily and quickly measured.

The area of the plane figure is measured by merely tracing the outline with the tracing point and figuring the result from the reading on the graduated wheel. This wheel is divided into 100 parts, each of which represents $\frac{1}{10}$ of a square inch, and each 10th can be read down to 100ths by the vernier on the instrument.

The simplest and cheapest planimeter measures up to 10 square inches and costs about $15. It can be bought of *Keuffel and Esser,* 101 Fulton Street, New York, or of the *L. E. Knott Apparatus Company,* Boston, Mass.

CHAPTER II

WHEN YOU BUILD YOUR HOUSE

You will find it a money saving deal to know something about building materials and how to choose and use them before you start in to build a house, or even a chicken coop.

Without such a working knowledge it is easy to pay high prices for poor grades and to use costly materials where cheaper kinds will do just as well. This is equally true whether you are going to do the job yourself or to hire someone to do it for you.

Comparative Cost of Buildings.—There are many kinds of materials used for building purposes, but the five chief ones are (1) *wood;* (2) *brick;* (3) *stone;* (4) *stucco;* and (5) *concrete.*

TABLE

Kind of Building	Cost
Wood	$5,000
Brick	5,575
Stucco	5,100
Concrete built with forms	5,600
Concrete built of blocks	4,200
Stone	5,600
Rubble	5,500

The comparative cost of buildings in which these materials are used varies in different localities, but the above table will serve to show them approximately.

Kinds of Materials to Use.—Where ordinary buildings are put up, the *piling,* if it is used, can be of wood or concrete. For *basement walls* to the first floor level, plank, brick, rubble, stone, concrete or hollow tile can be used.

Walls are built of wood, brick, stone, stucco, concrete, and occasionally of tile. *Chimneys* can be laid up of brick or built of concrete. All kinds of material, such as wood, asphalt and asbestos shingles, tin, galvanized iron, copper and zinc, slate and tile, are used for *roofing.*

Floors can be made of wood, concrete, tile, mosaic, rubber or pulp. The *outside trim,* such as doors and *finish,* windows and finish, pillars and turned work in general, and the inside *finish,* such as stairs, railings, ceiling beams, mantels, paneling, etc., all come under the head of *mill work* and can be bought ready made cheaper than you or a carpenter could possibly make them. They are far better, too, when bought.

Builders' hardware includes all kinds of hardware used on a building, and, finally, for *plastering,* wood and metal lath are used.

Now about Lumber.—*When the Tree is Felled.* —The word *timber* is used to mean both growing trees and cut trees and squared and sawed wood of the larger sizes, while the word *lumber* is taken to mean timber which has been sawed into scantlings and boards.

When a tree is sawed down, if you will look at the end of it, you will see in the center a little dot or circle, and this is called the *pith* of it. Around the pith there is a series of concentric rings called *annual rings*. The number of them shows the age of the tree, since there is a ring for every year of

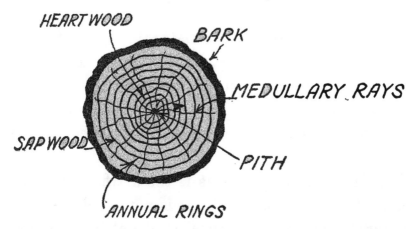

FIG. 10.—CROSS SECTION VIEW OF TREE SHOWING MEDULLARY RAYS AND ANNUAL RINGS

growth. Naturally, the size of the tree depends on the number of rings.

The wood next to the pith is called the *heartwood,* then comes the *sapwood* and finally the *bark.* The *medullary rays* are the lines that extend radially from the center to the circumference and all of which are shown in the cross section view, Fig. 10.

The Way Wood Is Seasoned.—When a tree is growing there is a large amount of sap in it. Since this is formed chiefly of water, when the tree is felled the water still remains in it. Before it can be used for building, the water must be dried out of it to some extent, and this process is called *seasoning.*

The two usual ways of getting rid of the water are by (1) *natural seasoning* and (2) *hot air seasoning.* If after the rough work has been done on a building it is left for a while before finishing, it dries out still more, and this is called *second seasoning.*

Natural Seasoning.—The natural way of seasoning lumber is the best way, but it takes a long time. It is done by piling it up so that the air can pass freely all around each piece. When you buy lumber for outside use, be sure to get it seasoned by this process.

Hot Air Seasoning.—This is the artificial method and, while it is quickly done, it is not nearly as good as natural seasoning. It consists of putting the lumber in a *drying room,* that is, a room which is kept hot by means of steam pipes. Wood seasoned in this way is very apt to shrink or swell with the changes of the weather. Hence it should never be used except for inside work.

How to Tell Good Lumber.—Trees have their diseases and parasites as well as human beings and in buying lumber, as in every day life, you must look out for them.

Lumber for building should be straight grained, be *clear,* that is, without knots, and be free from sap. You can always tell good lumber by its sweet smell. Its shavings will have a close-knit texture and a smooth, silk-like sheen. Don't buy lumber which has a bad smell and a chalky look.

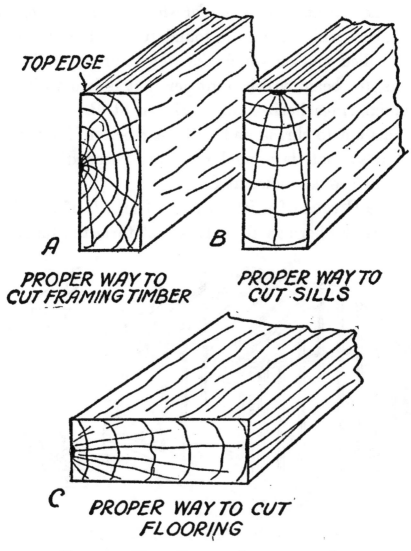

TOP EDGE

A

PROPER WAY TO
CUT FRAMING TIMBER

B

PROPER WAY TO
CUT SILLS

C

PROPER WAY TO CUT
FLOORING

FIG. 11.—HOW TIMBER SHOULD BE CUT

Using Lumber to the Best Advantage.—To
prevent lumber that is used for the frame of a build-

ing from shrinking, it should be cut so that the annual rings run in the same direction as the long end of the board, as shown at A in Fig. 11.

Where beams are used for *sills,* as the horizontal members which form the foundation of the building are called and on which the weight of the building rests, the beam will be stronger if it is laid with the annual rings horizontal, as shown at B. Flooring is less apt to shrink and will wear better if you can get it so that its annual rings are perpendicular to the surface, as shown at C.

The Frame of a Building.—The *sills* of a building are the horizontal timbers that form the founda-

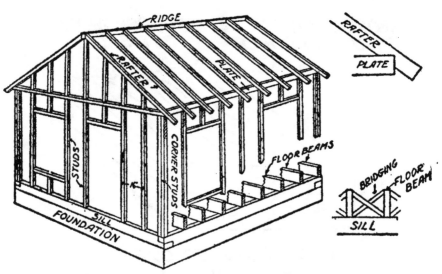

FIG. 12.—THE FRAME OF A BUILDING

tion on which the frame rests; the *studding* is the joists or upright posts in the frame; the *rafters* are the beams that give the slope to a roof, while the *weatherboards* are the outside boards of a building. These last are generally formed of *clapboards,* that is, boards whose lower edges are thicker than their upper edges, and they are nailed on so as to overlap and shed the rain. Fig. 12 shows the frame of a building.

Shingles are thin pieces of wood, or of other materials, usually 4 or more inches wide and 18 inches long, $\frac{1}{2}$ inch thick at one end and tapering down to $\frac{1}{8}$ inch thick at the other end. For the number and weight of shingles see Appendix I, and for the size, length and number of shingle nails to the pound see Appendix II. Finally, *finish* means the inside finish of a building and *trim* means the molding and finish outside.

Kinds of Woods for Building.—There are only about a dozen kinds of woods used for building purposes generally. These are named in the order of their relative costs, beginning with the cheapest. After each one is given its weight per foot in *board measure.* You can find the board feet either with a carpenter's steel square which is given under the subcaption of *Essex Board Measure* on page 38, or by a simple calculation.

Where to Use These Woods.—While the following kinds of woods are largely used in this part of the country (Eastern States) for the different structural parts of buildings, of course other woods can be used, if you find them cheaper or easier to get.

TABLE

Order of Relative Costs	Pounds per Foot of Wood, Board Measure
Hemlock is cheapest..............	2.08
Spruce.........................	2.30
Yellow pine.....................	3.17
White pine......................	2.30
North Carolina pine..............	
Beech..........................	
Chestnut.......................	3.12
Maple..........................	
Cypress........................	3.11
Oak............................	4.15
Cedar costs most................	2.97

For the sills use spruce or hemlock.

For the studding use pine or hemlock.

For the rafters use pine.

For the clapboards use North Carolina pine.

For the shingles get cedar, if possible.

And for finishing use yellow pine, spruce, cypress, maple, chestnut or oak, and use cedar for lining closets, if it does not hit your pocketbook too hard.

How to Preserve Wood.—To make wood last as long as possible it must be (1) thoroughly seasoned, (2) entirely free from cracks, or *shakes* as they are called, and (3) protected by some kind of a preservative.

There is never very much deterioration of *inside woodwork,* but it can be painted, oiled or varnished to advantage since, when it is so treated, it is more sanitary and sometimes more artistic.

As for *outside woodwork* on a building, the best

way to preserve it is to paint it. The best kind of paint is made of pure white lead and boiled linseed oil. Where wood is to be set in the ground, as posts, piles and flag poles, the ends can be *tarred, charred* or *creosoted.* [1] Tarring and creosoting are simple processes, for the wood needs only to be dipped into the former and soaked in the latter while it is hot. Charring is done by covering the end of the wood with charcoal and burning it.

Bricks and Brickwork.—A brick is a piece of molded clay which is dried in the sun and then burned in a kiln. Bricks come in two colors, *red* and *white.* The color of red bricks is caused by iron compounds in the clay, while light-colored bricks are made from clay which is practically free from iron.

Kinds of Bricks.—Bricks can be divided into two general classes, and these are (1) *stock* or *kiln-run* bricks, which are hard enough for the outside of buildings, and (2) *soft* or *salmon* bricks, which are used only for backing up and filling in.

There are a dozen grades of brick of the first kind and among these are (a) *common molded,* (b) *pressed* and (c) *enameled* bricks. There are half a dozen grades of the second kind and among these are common, soft and salmon brick.

The size of a standard brick in the United States is 2 x 4 x 8½ inches and its weight is about 4½ pounds.

---

[1] For wood preservatives write the *Carbolineum Wood Preserving Co.,* 36 Greene Street, New York, or the *Lyster Chemical Co.,* 61 Broadway, New York.

There are 66 cubic inches in a brick and, hence, it takes 26.2 bricks to make a cubic foot.

Bricks are very porous. A common brick will absorb as much as $\frac{1}{6}$ of its weight of water; but a really good brick should not absorb more than $\frac{1}{15}$ of its weight of water. To test a brick for porosity, weigh it, then let it soak in water over night and weigh it again. The difference in the weights will give the weight of the water absorbed.

Mortar for Brickwork.—In laying up a brick wall or chimney, the bricks are held together with a cement called *mortar,* which is made of *slaked lime* and *sand.*

Lime, or more properly *quicklime,* is a substance whose chemical name is *calcium oxide.* When it is mixed with water it generates a lot of heat and changes into *calcium hydroxide.* This process is known as *slaking.*

Sand is then mixed with it and the mortar thus made slowly absorbs *carbon dioxide* from the air which, acting on the *calcium hydroxide,* forms *calcium carbonate,* or *limestone,* and when the water dries out it becomes very hard. The purpose of the sand is to make the mortar porous so that the *carbon dioxide* can mix with it and it also prevents the mortar from cracking when it gets hard.

Plaster for Walls.—*Plaster* is simply mortar. Three different kinds of it are used for walls, and these are (1) *coarse stuff,* (2) *fine stuff* and (3) *gauged stuff.*

Coarse stuff is common mortar with hair mixed in it to bind it together. It is formed of 6 parts of lime, 12 of sand and 1 of hair. It is the first coat of plaster put on the lath, and the plasterer calls this *rendering*.

Fine stuff is made by mixing lime with water until it is about as thick as cream. After it has settled, the water is drained off. When the lime paste has hardened a little, a very small quantity of sand is mixed with it; it is then put over the coarse stuff, and this is called *floating*.

Gauged stuff is made by mixing 1 part of plaster of Paris with 4 parts of fine stuff. The plaster of Paris makes the stuff *set* very quickly, and so no more must be mixed at a time than you can put on before it gets hard. It is plastered over the fine stuff and is the last coat, or finish, and is called *setting*.

About Laying Brick.—In bricklaying a *course* is a continuous layer of bricks in a horizontal line, and a *bond* means the method used in laying the bricks in courses.

There are four chief bonds used in building brick structures, and these are (1) *common bond*, (2) *Flemish bond*, (3) *English bond* and (4) *cross bond*, all of which are shown in Fig. 13.

When you lay up a brick wall, the first thing to do is to have the foundation on which the courses are laid perfectly level. To find whether the top surface of the foundation and of the wall as you lay it are level, you must use a *level*, and to ascertain if the side of the wall is *plumb* an *upright level* or a *plumb* must be used. The construction of the level will be found in Chapter I. How to make square corners is also shown under the caption of *Tape Measures* in Chapter I.

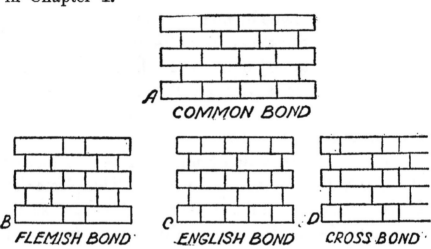

FIG. 13.—KINDS OF BONDS USED IN LAYING BRICKS

Measuring Brickwork.—The thickness of a brick wall is the number of bricks or half-bricks that it is built of. Brickwork is estimated by the *thousand*. The term *superficial foot* is used by masons and means square feet of surface. Walls of various thicknesses run like this:

TABLE

Thickness of Wall	Number of Bricks Thick	Number of Bricks per Superficial Foot
8¼ inches	1	14
12¾	1½	21
17	2	28
21½	2½	35

Stone and Stonework.—There are three kinds of stone used for building purposes, and these are (1) *field stone,* (2) *rubble* and (3) *cut stone.* They are laid either in (a) the *rough,* (b) in *ashlar,* or (c) in *courses,* as shown in Fig. 14.

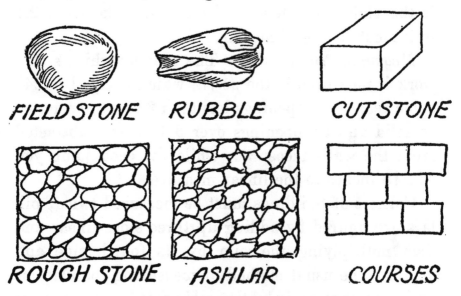

FIELD STONE   RUBBLE   CUT STONE

ROUGH STONE   ASHLAR   COURSES

FIG. 14.—KINDS OF STONE AND STONEWORK

By field stone is meant stones that are found on the surface of the ground, which are used just as they are picked up. Rubble is pieces of stone of all shapes and sizes as they come from the quarry, and cut stone is, of course, stone that is cut to shape and size in the quarry.

To lay field stone means to lay them in mortar or cement as they may fit best together. Ashlar is laid up in any order that the mason fancies, while cut stone is laid in courses.

Mortar for Stonework.—As the strength of stonework depends largely on the mortar that is used, it is better to use a *Portland cement* mortar than a mortar made of lime. A good cement mortar can be made by mixing 1 bag of Portland cement and 2 or 3 cubic feet of sand with enough water to give it the right consistency. This will make from 2.1 to 2.8 cubic feet of mortar.

Measuring Stonework.—The unit by which stonework is measured is the *perch,* which is equal to $24\frac{3}{4}$ cubic feet. All openings less than 3 feet are counted as solid and all openings over 3 feet are subtracted from the walls measured, while for each jamb you add 18 inches to the linear measure.

Corners of buildings must be measured twice; pillars are figured by adding up three sides linear and then multiplying the sum by its fourth side and depth. The usual method of measuring foundations and sizes of stone is by the cubic foot. *Base courses* and *water tables* are measured by lineal feet; sills and ashlar are measured by superficial feet.

Stucco for Buildings.—Stucco is simply a mortar made of Portland cement, sand, lime and water and when rightly made it is enduring as the ages. It is used as a plaster for the outside walls of buildings and makes a beautiful fire-resisting structure built at a low cost and with no expense for upkeep.

Ways of Using Stucco.—There are three ways of applying stucco and these are (1) on *wood sheathing,* (2) on *ribbed metal lath* and (3) on brick, stone, tile and cement blocks. Where sheathing is used it is covered with sheathing paper, then *furring strips* are put on upright over it, and either wood or ordinary metal lath is nailed across the furring strips, as shown at A and B in Fig. 15.

Where sheathing is not used, *ribbed* metal lath is nailed on the studding direct, with the ribs inward, and the stucco is plastered on both the front and

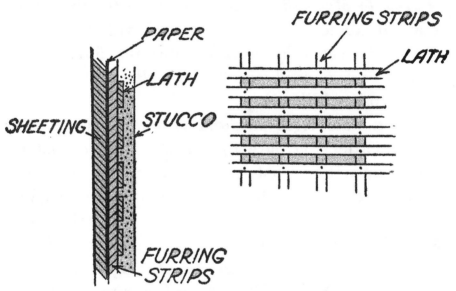

FIG. 15.—HOW STUCCO IS PUT ON

back of it until it is about 2 inches thick. When it sets you will have a wall as hard as adamant of reën-forced Portland cement mortar. The tools needed are shown in Fig. 16. Brick, stone, tile or cement can also be given a coat of stucco, but the surface must be rough enough to make it *key,* that is to stick tight.

Putting on Stucco Mortar.—Three coats of stucco mortar must be put on to make a good job. The first coat, which is put on the face of the lath, and the second coat, which is put on the back of the lath, should each be $\frac{3}{8}$ inch thick, while the last and finishing coat should be $\frac{1}{4}$ inch thick.

When stucco is put on ribbed lath, the first, front and back coats should be from $\frac{3}{8}$ to $\frac{1}{2}$ inch thick, and the finish coat $\frac{1}{4}$ inch thick.

How to Make Stucco Mortar.—For the first two coats of stucco mix 3 parts of sand with 1 part of Portland cement by *volume*. For the finish coat mix $2\frac{1}{2}$ parts of sand with 1 part of Portland cement and $\frac{1}{10}$ part of lime.

Use a water-tight platform to mix the stucco on so that, after you have the right amount of water for mixing, it will not leak away. Sometimes hair or fiber is used for the first coat of stucco, as in ordinary mortar. If either is used, it is mixed in after the mortar is made. Mix the mortar until it is smooth and is of the same color throughout.[2]

Building with Concrete.—Concrete is *your* building material, by which I mean that you can build any ordinary structure of it with the help of common labor.

It is timeproof, waterproof and fireproof and, though it costs a little more than wood in the first place, it does not cost anything for paint and repairs after it is built. It is different from brick

[2] If you are interested in building a stucco home, a garage or a barn, write the *Atlas Portland Cement Co.*, 30 Broad Street, New York, and they will send you plans and specifications without cost.

and stone in that you can always get the materials to make concrete wherever you live.

What Concrete Is.—Concrete is made up of four materials and these are (1) *Portland cement,* (2) sand, (3) stone or gravel and (4) water. It is called Portland cement because it is about the same color as the limestone quarried on the Isle of Portland, England.

It is made by heating limestone, clay and sand, or blast furnace slag, until they are changed into a powder and when this is mixed with water it will set hard and water will not affect it in any way. Portland cement is manufactured in great mills where it is packed in bags which hold about 1 cubic foot each. It is then shipped to the four quarters of the globe, so you will have no trouble in buying it wherever you are.

Materials for Concrete.—*Testing Portland Cement.*—Before the cement is used it must be kept perfectly dry or it will absorb moisture and get hard.

Sometimes when bags of cement are piled on each other, the cement will *cake,* but this does not injure it in any way. To test cement that is lumpy, pinch a piece of it between your fingers and see if it will break up; if it will not, it is useless for concrete.

Testing Sand.—Sand, or *fine aggregate,* as it is called, must not have any loam, clay or other impurities in it. The particles that form it must not be too large to pass through a sieve with $\frac{1}{4}$ inch meshes.

To test sand for impurities, take a little while it

is still moist from where it is dug and rub it between the palms of your hands. If it does not soil them it is free enough from loam to use, but if it does, it must be washed by shoveling it onto a screen and washing it down with water.

Crushed Stone or Gravel.—Either gravel or crushed stone, or *coarse aggregate,* as it is called, can be used for concrete. It must be clean, free from impurities, and should not be less than $\frac{1}{4}$ inch in size and never more than half the thickness of the concrete you are placing. Finally, well water is the best kind to use for making concrete.

Mixtures of Concrete and Where to Use Them.

—The following mixtures are largely used and will give satisfaction for the purposes named.

A Rich Mixture.—Use 1 part of cement, $1\frac{1}{2}$ parts of sand and 3 parts of coarse aggregate; this makes a good cement for waterproof buildings and roads.

The Standard Mixture.—Use 1 part of cement, 2 parts of sand and 4 parts of coarse aggregate. Useful for floors, roofs, tanks, conduits, sewers and reën-forced work.

A Medium Mixture.—Use 1 part of cement, $2\frac{1}{2}$ parts of sand and 5 parts of coarse aggregate. Largely used for foundations, piers, walls, etc.

A Lean Mixture.—Use 1 part of cement, 3 parts of sand and 6 parts of coarse aggregate. Good for backing stone masonry, massive concrete work and large foundations.

Mixing Concrete.—The materials of which concrete is made can be mixed either (1) by *hand,* or (2) by *machine.* It should be mixed close to the place where you are going to use it; otherwise it will set before you can place it. For ordinary work it should be about as thick as jelly, and it should be mixed just as mortar is.

Placing Concrete.—There are two ways to use

WIRE NETTING
¼"MESH

SCREEN FOR SAND

←WOOD

FLOAT FOR FINISHING
OFF CONCRETE

Fig. 16.—The Only Tools You Need for Concrete Work

concrete for building and these are (1) to mold it in *forms,* and (2) to cast it in *blocks.*

To make a form for a wall, build up two sides of boards 1 inch thick and brace them so that the space between them is as thick as you want the wall, as shown at A in Fig. 17. The way to make forms for a pier and for steps is shown at B and C.

Rub soap or crude oil on the inside of the form and pour the concrete mixture into it. It will take from two days to a week for the concrete to set hard and then you can take off the form.

Concrete blocks, as shown in Fig. 18, are molded

either hollow or in solid veneer and they are easy to make and set. If you are interested in building with them, write to the *Ideal Concrete Machinery Company* of South Bend, Indiana, for a free booklet of their machines and equipment.

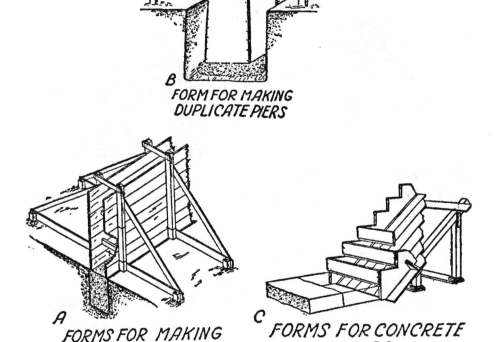

B
FORM FOR MAKING
DUPLICATE PIERS

A
FORMS FOR MAKING
A WALL

C
FORMS FOR CONCRETE
STEPS

FIG. 17.—FORMS FOR PLACING CONCRETE

Finishing Concrete Surfaces.—Ordinary concrete work does not have to be finished, but you can improve the surfaces of walls by rubbing them with a *cement mortar brick,* made of 1 part of cement and 2 parts of sand, and keeping it flushed with water while you are doing it. Designs for forms of all kinds can be had for the asking by writing to the

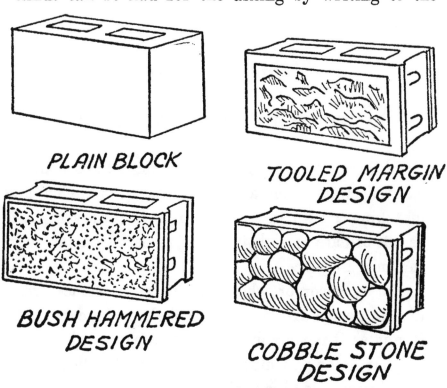

PLAIN BLOCK

TOOLED MARGIN DESIGN

BUSH HAMMERED DESIGN

COBBLE STONE DESIGN

Fig. 18—Some Concrete Block Designs

Atlas Portland Cement Company, 30 Broad Street, New York.

CHAPTER III

A WATER SYSTEM FOR YOUR PLACE

In these days of power and pumps, the scheme of carrying water from a well to supply the kitchen and of taking a bath in a washtub on Saturday night is as out-of-date and about twice as barbarous as cooking in a fireplace.

But however or wherever the water comes from, disease germs are more than likely to be carried by it, and as it is your first duty to safeguard the health of your home you must know to a certainty that the water supply is absolutely pure.

Kinds of Water Supplies.—There are three kinds of *water supplies,* or places from which to get water, and these are (1) *surface,* (2) *shallow underground* and (3) *deep underground* supplies.

The surface supplies are the ponds, streams, rivers and cisterns and all of these are very apt to be polluted with disease germs. This untoward condition is largely due to contamination from sewage, that is, the sewage is either emptied into them or else seeps into them from nearby sources. Whatever you do, don't use water from a surface supply for drinking, or cooking, or even washing dishes, unless it has been thoroughly purified first.

The water of shallow wells is also often disease bearing, but deep wells are very seldom so. In any event, remember that water which looks perfectly clear may have disease germs in it.

How to Purify Water.—*By Boiling.*—A simple and sure way to get rid of all the germs in water is to boil it; but it is not enough to merely bring the water to a boil, for a typhoid germ is as immune to heat as an asbestos cat. Boiling the water for 15 minutes or so will kill most of the germs, but to be sure that all of them are killed the water must be boiled *twice*.

By Filtration.—A great deal of impure matter in water can be removed from it by *filtering,* that is, by straining it through some kind of porous material. Filters that are made to screw on to the faucet remove some of the impurities, but most of the germs go on through.

Filters made of charcoal, sand and gravel [1] remove nearly all the impurities, but still some of the germs get through. By adding a very small amount of alum to the water the impurities and nearly all the germs will stick to the particles of it which then fall to the bottom, or are *precipitated,* as it is called.

---

[1] A complete description of a cheap and good filter of this kind, with drawings, is given in my "Home Handy Book," published by D. Appleton and Company, New York.

The *Pasteur filter*[2] is a good one for the household. The water flows in through the top and its weight forces it through an unglazed porcelain cylinder, the top end of which is closed. To make the filter effective the cylinder must be taken out

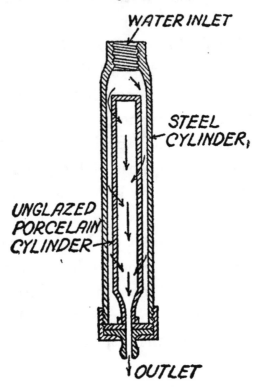

WATER INLET

STEEL CYLINDER,

UNGLAZED PORCELAIN CYLINDER

OUTLET

Fig. 19.—The Pasteur Water Filter

every day and the mud and slime scrubbed off with a brush. Otherwise it will form a breeding place for the germs instead of purifying the water. It is shown in Fig. 19.

By Distillation.—To distill water on a large scale

[2] Sold by the Consolidated Filters Co., 136 West 65th Street, New York.

requires a costly apparatus, but a small *still* can be easily made that will distill enough drinking water for the family.

The still is formed of (1) a *boiler* holding a couple of gallons of water, which sets on a stove, and (2) a *condenser* hung from the ceiling; a pipe connects the boiler and the condenser and carries the steam from the former to the latter. The condenser is made of an inverted funnel with a large pipe soldered to the mouth of it, while around the funnel is a vessel filled with water.

The lower end of the pipe is closed and a faucet leads from it to a covered bucket. The construction of the still is shown in Fig. 20.

The still should be made of heavily tinned copper, and no solder should be used on the inside of the seams. Now when the steam passes into the condenser from the boiler, it strikes the funnel and the cold water which surrounds it condenses it when it trickles down the large pipe and can be drawn off into the bucket as it is required.

The Amount of Water Needed.—The amount of water used will, of course, depend on the size of the family and, if you live on a farm, on the kind and number of stock you have.

It takes on an average of from 25 to 40 gallons of water a day to keep each member of the family supplied with enough to drink, to cook with and to bathe in; hence a water supply for a family of five

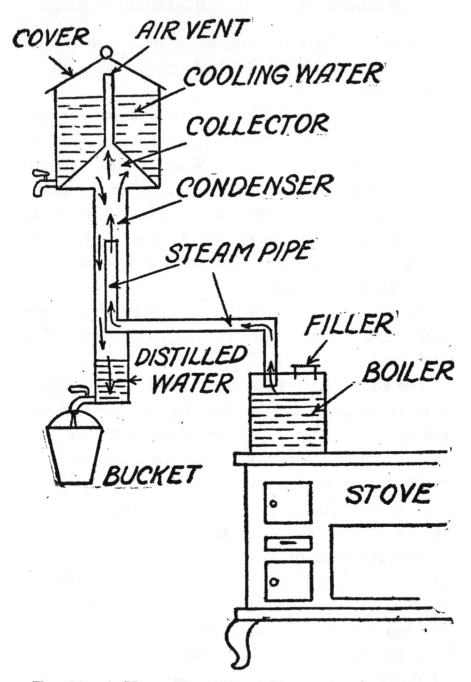

COVER — AIR VENT — COOLING WATER — COLLECTOR — CONDENSER — STEAM PIPE — FILLER — BOILER — DISTILLED WATER — BUCKET — STOVE

FIG. 20.—A HOME-MADE WATER DISTILLING APPARATUS

or six should have a tank, if one is used, with a capacity of something over 200 gallons.

Where there is stock, each cow needs about 12 gallons; each horse about 10 gallons; each hog about 2½ gallons; each sheep about 2 gallons, and there must be a small surplus for the dog and the cat. If you intend to sprinkle the lawn and the garden and have fire protection, allowance must also be made for an additional supply.

Schemes for a Water Supply.—There are three schemes in general use by which you can have running water in your house and on your farm and these are (1) the *gravity* system, (2) the *air pressure* or *pneumatic* system and (3) the *automatic air pressure* or *auto-pneumatic* system.

The Gravity System.—In this system the water is pumped either by hand or power into a tank set as high as possible; this is usually in the attic, as shown in Fig. 21, or on the tower of a windmill. The tank can be of wood or steel and either in the shape of a cylinder or a rectangle. Wood tanks should be made of cedar or cypress and these can be lined with tinned copper, but lead must not be used.

The Air Pressure or Pneumatic System.—In this system an air-tight steel tank is set in the basement, or in an underground vault,[3] and it is connected with the cistern or well by a force pump.

The water is then pumped into the tank against

[3] This keeps it cool in summer and prevents it from freezing in the winter.

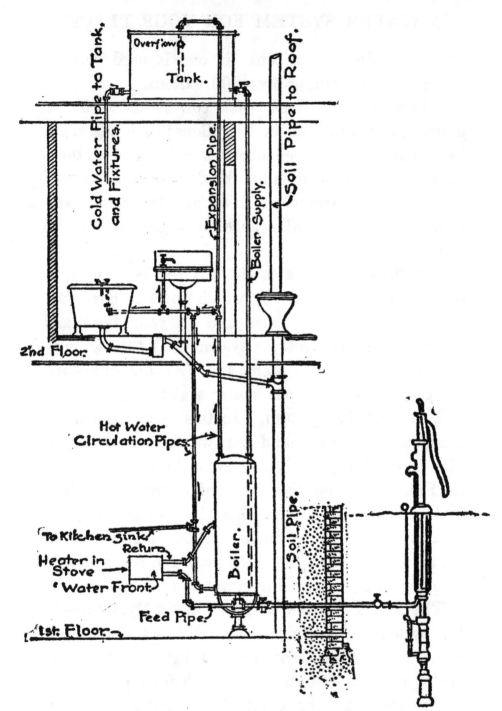

FIG. 21.—A GRAVITY WATER SYSTEM

the air that is in it. This compresses the air, and
the pressure set up will force the water through the
pipes to a height of a hundred feet or so. The tank
is fitted with a *water gauge* and an air *pressure*

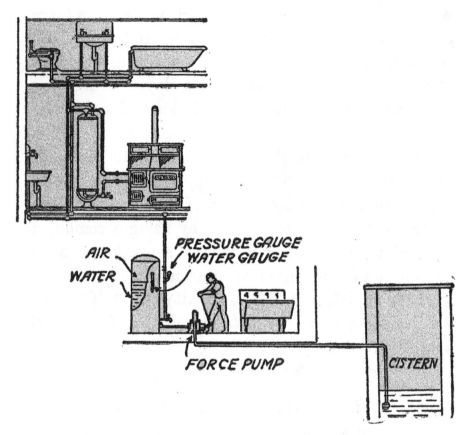

FIG. 22.—THE HYDRO-PNEUMATIC SYSTEM

gauge,[4] so that you can see at a glance the amount
of water there is in the tank and what the air pressure in pounds in it is. The outfit is shown in Fig.
22.

[4] A description of both of these gauges will be found in
Chapter V.

The tank can be set up on end, that is, upright, or lengthwise, that is, in a horizontal position, according to the room you have. The size of the tank will, as before, depend, of course, on the amount of water needed. A 220 gallon tank is about as large as you can use to advantage with a hand pump, and this will supply a family of five or six, provided all of them do not take a bath every day. In figuring the size of the tank, allow $\frac{1}{8}$ of the space for the compressed air.

As water absorbs the compressed air in the tank, means must be provided to supply air to the tank. This is done either by (1) an air inlet valve in the suction pipe of the pump, (2) by using a combined air and water pump or (3) by a separate air compressor run by an engine or other motive power.

How to Figure the Capacity of a Tank.— *To find the quantity of water a cylindrical tank will hold,* figure it this way:

$$C = D_2 \times 0.7854 \times d \times 7.48$$

where C is the capacity in gallons of the tank you want to find,

D_2 is the diameter of the tank in feet squared,

0.7854 is a constant,

d is the depth of the tank in feet, and

7.48 is the number of gallons in a cubic foot.

To find the quantity of water a rectangular tank will hold, use this formula:

$$C = L \times W \times D \times 7.48$$

where C is the capacity in gallons of the tank
which you want to find,

L is the length of the tank,

W is the width of the tank,

D is the depth of the tank, and

7.48 is the number of gallons in a cubic foot.

The Weight of Water.—In putting up a tank,
due consideration must be given to its weight on the
structure supporting it, when it is full of water.
Knowing that the weight of a gallon of water is 8.4
pounds and that a cubic foot of water weighs 62.5
pounds, it is easy to find the total weight of water in
either a cylindrical or a rectangular tank.

The Automatic Air, or Auto-pneumatic System.—As its name indicates, this system is worked
by compressed air which automatically delivers the
water direct from a lake or river, cistern or well, to
the faucets where it is to be used. The water, of
course, must be free from dirt.

The apparatus consists of (1) an *engine* or motive power of some kind, (2) an *air compressor,*
(3) a steel *air tank* and (4) an *auto-pneumatic water pump.* The engine drives the compressor which
pumps the tank full of air to a pressure of from 40
to 100 pounds per square inch. The air tank is connected directly with a pipe line to the pump, which
is placed near the bottom of the well or cistern.

Since the air in the air tank is under a high pressure and the water pump works on a low pressure, a *reducing valve* is placed in the pipe line to lower the pressure of the air and make it flow in a steady stream to the pump.

The pump is the chief part of the outfit and is formed of two steel cylinders. These are connected at the upper ends to the compressed air tank. In the bottom of each cylinder is an inlet valve for the water

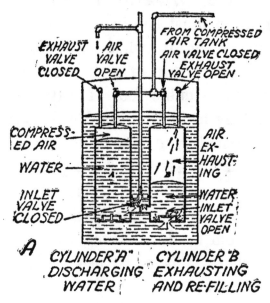

FIG. 23A.—THE AUTO-PNEUMATIC WATER PUMP

to flow from the well or cistern, as in any force pump. Each cylinder is also fitted with an air exhaust valve and, when the pump is submerged in the water, the pipes from the exhaust valves project above the surface of the water. Finally, each cylinder is connected to the delivery pipe which carries the water to the faucets. The operation of the system will be readily understood from Fig. 23.

About Pumps and Pumping.—*Kinds of Pumps.*—There are three kinds of pumps that are used for home and farm pumping and these are (1) the *lift* or *suction* pump, (2) the *force* pump and (3) the *centrifugal* pump. The lift pump is usually worked

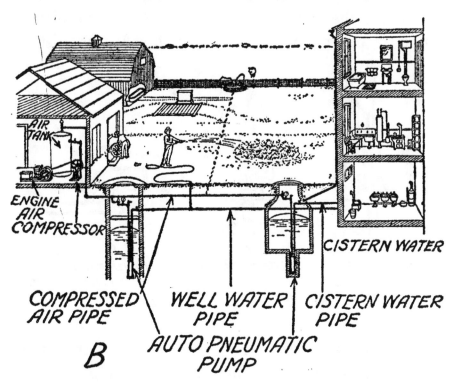

FIG. 23B.—THE AUTO-PNEUMATIC WATER SYSTEM

by hand; the force pump is worked either by hand or power; and the centrifugal pump is usually operated by power.

A *lift pump,* of which a cross section is shown at A in Fig. 24, consists of a cylinder, a piston, a couple of valves and a suction pipe whose lower end dips below the level of the water in the cistern or well.

When the piston is worked, the air from the pipe is pumped out and then the air pressing on the surface of the water pushes it up through the pipe and through the lower valve into the barrel.

When the piston moves down again, the lower valve closes and the water in the cylinder opens the piston valve as the piston sinks below it. As the piston

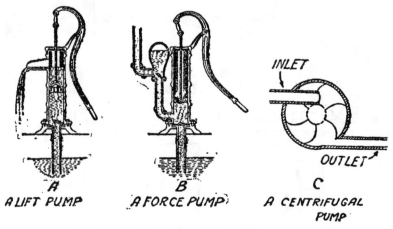

A LIFT PUMP   A FORCE PUMP   A CENTRIFUGAL PUMP

FIG. 24.—KINDS OF PUMPS

is again raised it lifts the water on top of it to the spout, and, at the same time, the pressure of the air forces more water up through the suction pipe.

A *force pump,* shown at B, is usually made with a solid piston. The upper valve is set in the outlet pipe which opens below the piston. When the piston moves up, water is drawn up into the cylinder by atmospheric pressure; when it moves down, the valve in the suction pipe is closed and the water is forced through the upper valve into the discharge pipe.

When the piston is raised again, the valve in the

outlet pipe is closed so that the water above cannot flow back. At the same time the pressure of the air forces more water from the well into the cylinder.

A *centrifugal pump* is a rotary pump, that is, it consists of a number of curved blades fixed to and radiating from a shaft, like the spokes of a wheel from the hub.

These blades slide against the sides and the inside rim of the pump case. This prevents the water from leaking between the blades and the case when it is pumping. The intake water pipe is placed in or near the center of the case and, as the water flows into it, the swiftly revolving blades throw it out by centrifugal force into the delivery pipe. It is shown in cross section at C.

The Action of Pumps.—*The Lift Pump.*—A lift pump will only lift water effectively about 20 feet because it depends on atmospheric pressure, and hence the cylinder of a lift pump must not be set higher than this distance above the level of the water in the well or cistern.

A good way to get rid of the suction lift is to have the cylinder close to the water, or submerge it, if this can be done, as this keeps the pump primed all the time. As the lift of the water above the piston does not depend on atmospheric pressure, a pump of this kind can be used for greater depths. Pump cylinders are made which will go into wells as small as 2 inches in diameter.

The Force Pump.—Water can be raised to any

height by means of a force pump. The purpose of the *air chamber* on a force pump is to make the water flow in a continuous stream through the delivery pipe.

In this case, when the water is forced into the air chamber, it covers the mouth of the delivery pipe and, as it rises, it compresses the air that is in the chamber. The pressure of it soon becomes great enough to force the water through the delivery pipe in a steady stream.

The Centrifugal Pump.—As the blades of a centrifugal pump do not fit air-tight, it is not positive in its action, like a valve pump. It will not, therefore, exhaust the air from the suction line, so it must be primed every time before it is started, no matter how small the suction is.

To prime the pump, it must be at rest and both the suction pipe and the pump case must be filled with water. A small centrifugal pump will then only lift water 10 or 15 feet, but it will deliver it to a height of 35 feet or so. These pumps are largely used for pumping water to boilers in steam heating plants.

To Prevent Pipes from Freezing.—Where a water pipe extends above the ground or is above the frost level, it must be protected from freezing and this can be done with a *frost box*.

To make this covering build three box tubes around the pipe, as long as the part of it you want to protect, and cover the outside of each one with tar paper. Have an air space of 6 inches all round between the inside and the pipe and an air space of 2 inches between each of the other two boxes. Keep the boxes separated from each other by blocks or *trimmers,* and you will have a good insulation against the cold. Ordinary pipe coverings will not keep water pipes that are out of doors from freezing.

To prevent underground water pipes from freezing, the pipes must be buried to a depth of 3 or 4 feet.

When a Water Pipe Is Frozen.—Where an exposed pipe freezes, wrap woolen cloths around the frozen part and pour hot water on it until it thaws. If an underground pipe freezes, you will have to dig down to it and thaw it out. If it is a large pipe, you can do this by building a fire around it. If a lead pipe bursts, it can be soldered, but if an iron pipe bursts a new length of pipe will have to be put in.

A Word on Plumbing and Sewage.—*Plumbing.*—It is easy to do your own plumbing, for the days of lead pipe with the trouble of making *wiped joints* are over; instead iron pipe in all sizes and with all the necessary fittings can be bought ready to put together.[5]

Use 1 or 1¼ inch pipe for the main piping to the supply tanks. The table on the next page shows the sizes of pipes required for various branches.

Make all joints and fittings water-tight with *red*

---

[5] Write Sears, Roebuck and Co. for their catalogue on *Plumbing.*

TABLE

Branches from Main Pipe	Inches	Branches from Main Pipe	Inches
To basin cocks......	½	To water closet flush pipes......	1¼ to 1½
" bath "	¾ to 1	For kitchen sinks..	⅝ to ¾
" water closet flush tank..........	½	For pantry sinks..	½
" water closet flush valve.........	1 to 1¼	To slop sinks......	⅝ to ¾

lead. Have a drain cock at the lowest point in the system so that you can let out the water in cold weather to prevent the pipes from freezing and bursting. Also keep the hot water pipes far enough away from the cold pipes to prevent an exchange of heat between them.

Sewage.—Sewage must be disposed of by a *septic tank system.* In this system the sewage empties in an air-tight tank which is connected to a second tank by an overflow pipe. The first tank is large enough so that it does not overflow for about 12 hours. The germs formed in the sewage in this air-tight tank partly decompose it when it is discharged into the second tank, where it undergoes a like purification. It can then be discharged to the outside air free from obnoxious odors.

CHAPTER IV

A HEATING PLANT FOR YOUR HOME

There are many kinds of forces that perform amazing feats, and one of the most active of these is *heat*.

What Heat Is.—The most common way of producing heat is by burning something, or *combustion,* as it is called. Combustion is caused by chemical action.

As an illustration take *oxygen* and *carbon.* These two substances have a great attraction for each other, and, if you can get a large quantity of oxygen and a lot of carbon stored up separately, you have the means for making a fire and hence of generating heat.

Now air is formed of $\frac{1}{5}$ part by volume of oxygen, so you always have a supply of this gas at hand. As coal is nearly pure carbon, you can get a supply of this (sometimes) if you have the money. Here, then, are your separate stores of these combustible chemicals, and all you need to do to start the chemical action of combustion is to ignite them.

When combustion is going on, the particles, or *molecules,* as they are called, of oxygen and carbon combine and they vibrate at a rapid rate. These rapid to and fro motions impinging on our sense of feeling set up the sensation that we call heat.

What Temperature Means.—Heat and *tempera-*

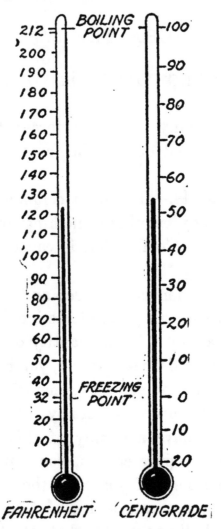

FIG. 25.—FAHRENHEIT AND CENTIGRADE SCALES COMPARED

ture mean two entirely different things, though they are very closely related.

Temperature is not only the *degree* to which a

body is heated but, in *physics,* it is defined as *that property of a body which determines the transfer of its heat to some other body.*

When a body gives out heat, its temperature falls and, conversely, when a body receives heat its temperature rises. Temperature is measured by *thermometers,* and these are *graduated* in different ways. On all of them, however, there are two fixed points, namely, (1) the *freezing point* and (2) the *boiling point.*

The *Fahrenheit* thermometer scale is the one used in this country for all ordinary temperature measurements and the *centigrade* scale is used for all scientific measurements. Both of these scales are shown in Fig. 25, and are marked like this:

Minimum and maximum points	Fahrenheit	Centigrade
Freezing point.................	32 degrees	0 degrees
Boiling point...................	212 degrees	100 degrees

How Heat Warms a Room.—When you have a body that is heated to a higher temperature than another body near to it, the heat from the warm body always passes to the cool body until the heat, and consequently the temperature, of both are the same.

Thus if a room is heated to a temperature of 68 degrees Fahrenheit and the air outside is colder, the heat will leak through the windows and walls and, unless more heat is constantly supplied to the room, its temperature will fall.

How Heat Is Measured.—Since heat is a force, it can be measured quite as exactly as wind, water, steam or electricity.

Just as the unit of English lineal measure is the inch and the unit of weight is the pound, so also the *unit of heat* is the *British thermal unit,* or B. T. U., as it is called for short, and this is the amount of heat that is needed to raise the temperature of 1 pound of water 1 degree Fahrenheit.

About Heating and Ventilating.—On first thought heating and *ventilating* may seem to have little in common, yet a supply of pure, fresh air is even more necessary than a supply of hot air. But when you get both of them together you have the ideal conditions that make for health and comfort.

Since this is true, in planning a house you should provide for ventilating it at the same time that you consider the best way of heating it, for the air supply should be heated before it is admitted to the rooms.

Kinds of Heating Plants.—There are seven ways a building can be heated, and named, these are (1) by *fireplaces,* (2) by *stoves,* (3) by *hot air furnaces* (4) by *hot water systems,* (5) by *steam heating plants,* (6) by *gas burners* and (7) by *electric heating apparatus.*

The Cozy Fireplace.—Next to a fire in the center of a wigwam with a hole in its top for the smoke to get out, the fireplace is the oldest scheme of man to heat his abode.

Fireplaces are used in present day homes chiefly for the cheer and comfort they offer. They are very wasteful of fuel, for 85 per cent of the stored up energy of the wood or coal goes up the chimney; but they are good ventilators, for the chimney makes a draft and this pulls fresh air into the room.

To get the best results, fireplaces should be lined with brick, and where grates are used they should be set well above the level of the floor.

The Cheap Old Stove.—Since a stove is cheap, portable and far more efficient than a fireplace, it has all but supplanted the latter in the cheaper houses. The bad features about a stove are that it dries the moisture out of a room, it does not draw in any fresh air, and it makes a lot of dirt. The best thing that can be said of a stove is that it uses 50 per cent or more of the heating energy of the fuel it burns.

The Hot Air Furnace.—To get rid of the trouble and dirt of fireplaces and stoves, some genius got up the scheme of hot air heating.

In this system a furnace is used which has a double shell; the fire is built on a grate in the inside one, or *fire pot,* and air is drawn in between it and the outer shell or casing, where it is heated. The hot air then passes through large pipes in the top of the furnace, or *dome,* as shown in Fig. 26, where it flows into the various rooms through iron openings, called *registers,* set in the floors or baseboards. The registers are fitted with valves by which the heat can be turned on or cut off.

Heating detached houses by hot air is seldom satisfactory, for it is always hard to heat the rooms on the side against which the wind blows. This can

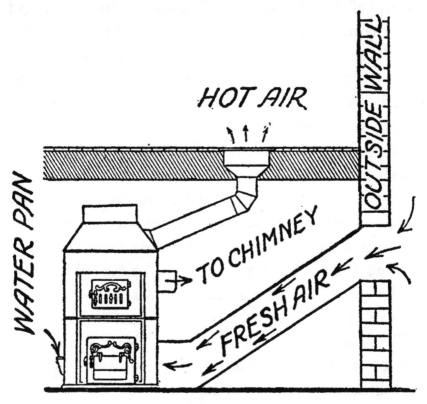

FIG. 26.—HOW A HOT AIR FURNACE WORKS

often be remedied by installing a fan in the pipe leading to the cold room. If you are putting in a hot air system, be sure you use a furnace big enough and that the pipes are large enough. Hot air systems are cheaper to install than hot water or steam systems, and they are very simple to operate.

A HEATING PLANT FOR YOUR HOME

The Hot Water System.—There are several kinds of hot water heating systems, but the *gravity* or *low pressure* system is generally used for heating homes.

There are two chief kinds of low pressure systems and these are (1) the *one pipe* system and (2) the

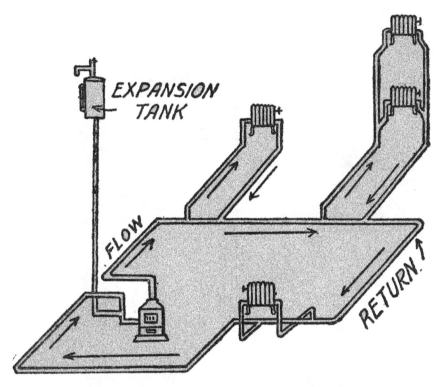

Fig. 27.—A One Pipe Hot Water System

two pipe system. The one pipe system is the cheapest to install and it will give you very good service. The two pipe system takes twice as much piping and more labor to put in and besides it is apt to *short circuit* unless the work is well done.

Either kind is easy to take care of, and they make it possible to regulate the temperature. A hot water system costs more than a steam heating plant, because the pipes and the radiators must be larger, but it is more economical in fuel consumption.

In putting in a one pipe system, have the piping all of one size. The horizontal supply pipes that

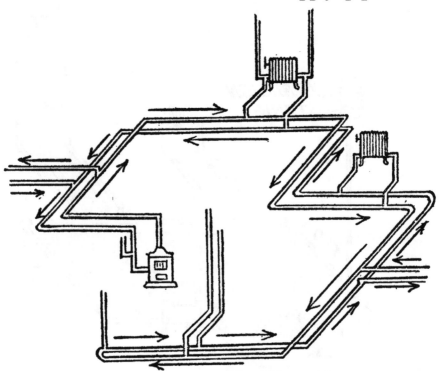

Fig. 28.—A Two Pipe Hot Water System

branch off from the main circuit, as shown in Fig. 27, should be short, or, if they have to be long, then give them as much pitch as you can. A two pipe system is shown in Fig. 28.

Steam Heating Plants.—There are several kinds of steam heating systems, but the one that is in general use in this country is the *low pressure live steam* system.

A low pressure steam boiler is made just about the same as a hot water boiler, but it is fitted with a steam gauge, water gauge and safety valve which operates the damper to regulate it.

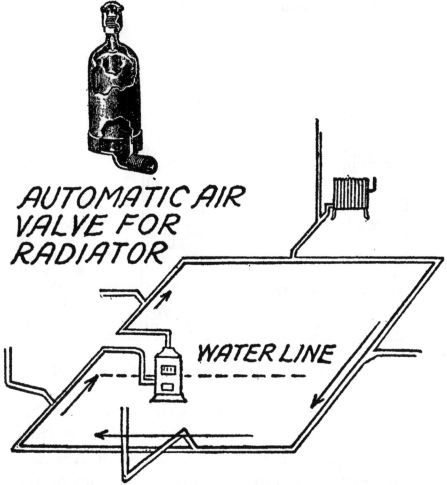

FIG. 29.—A ONE PIPE STEAM HEATING SYSTEM

As in hot water heating, there are two schemes of piping used, namely, (1) the *one pipe system* and (2) the *two pipe system*. In the one pipe system, and this is the one that is most widely used, a single

pipe leads off from the main line and is connected to
the bottom of the radiator which has a small automatic valve on the opposite side, as shown in Fig.
29.   The pressure of steam is usually only about 4
or 5 pounds per square inch, and this is measured by
a *steam gauge.*

The main pipe line should have a fall of 1 inch

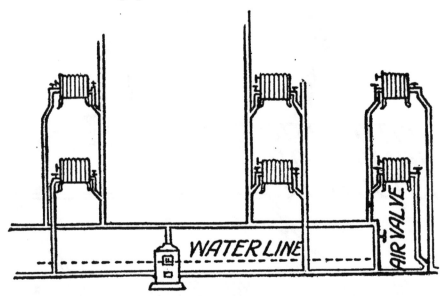

FIG. 30.—A TWO PIPE STEAM HEATING SYSTEM

in every 10 feet, and the pitch should be such that
the live steam and the water which condenses from
it will flow through the pipe in the same direction.
An automatic air valve is fitted in the top of the
riser to let the air that collects in the main line pipe
escape.

In this system the radiators are connected on one
side with the main line of pipe and on the other
with the return pipe, as shown in Fig. 29.   Where

this system is used the pipes can be of smaller diameter than with the one pipe system. The lower part of the pipe line of a steam heating plant should always be even. Fig. 30 shows a two pipe line steam heating system.

Noise in Steam Pipes.—The reason steam pipes crack and pound is because the hot steam strikes on the cold pipe or water caught in traps or pockets in the pipe. When it is the latter, the cold water condenses the steam and this forms a vacuum which pulls the water toward it with great force.

The chief thing you want to look after in installing a steam heating plant is to give the pipes sufficient pitch to carry off all the water formed in them and not to have any uneven places or pockets to catch the water.

To Find the Size of Heater Needed.—The size of a hot air furnace, a hot water heater or a steam boiler needed to heat a house can be found in several ways.

The simplest method depends on the cubic contents of the building to be heated and this is found by multiplying the total exposure of the house by 50 and dividing it by 30,000 and the result will be the number of square feet needed for the grate.

Gas Heaters.—Gas is used in nearly every city for cooking ranges, but it is used only to heat houses with as a makeshift, except in districts where there is natural gas.

Electric Heating Apparatus.—Where water

power is to be had, as at Niagara Falls and Boise, Idaho, electricity can be generated cheaply enough to be used to heat houses. But where fuel must be burned to generate it, electric heating of homes is all but out of the question.

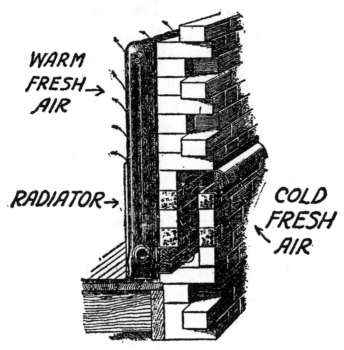

FIG. 31.—HOW TO GET GOOD VENTILATION

How to Get Good Ventilation.—Should you plan to build a house and intend to use either hot water or steam to heat it, you can have ventilating ducts put in the walls so that the fresh outside air will pass between the columns of the radiators and reach the room in a heated condition. The scheme is shown in Fig. 31.

CHAPTER V

HOW MACHINES ARE MADE AND USED

When you look at a complicated machine it hardly seems on first sight to be built up of just two simple mechanical principles, or *powers,* as they are called, but this is, nevertheless, true.

How Machines Are Made.—These two powers are (a) the *lever* and (b) the *inclined plane,* and

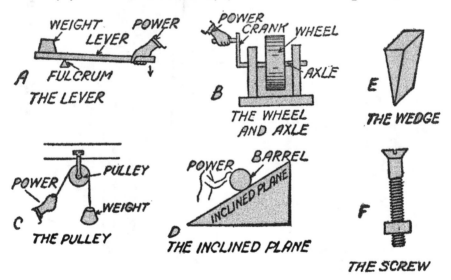

FIG. 32.—THE SIX SIMPLE MACHINES

both have been so improved that, including them, six *fundamental mechanical movements,* or *simple machines,* as they are termed in physics, are the re-

75

sult. The names of these simple machines are (1) the *lever,* (2) the *wheel and axle,* (3) the *pulley,*

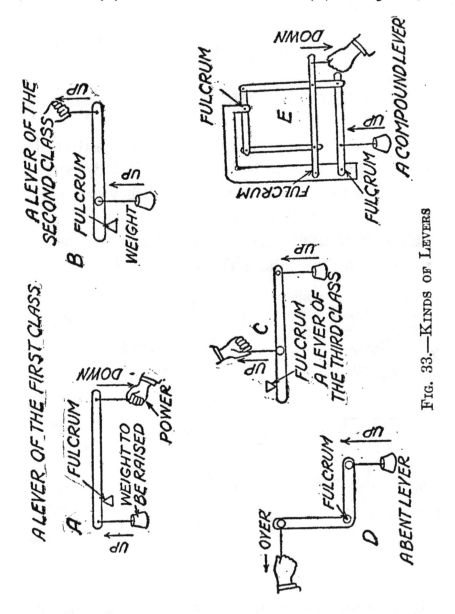

FIG. 33.—KINDS OF LEVERS

(4) the *inclined plane,* (5) the *wedge* and (6) the *screw.* The wheel and axle and the pulley are only

advanced forms of the lever, while the wedge and the screw are higher forms of the inclined plane, as you will presently see. All of these are shown in Fig. 32, the different kinds of levers in Fig. 33 and the various kinds of pulleys in Fig. 34.

From the above six simple machines a large number of *mechanical movements* have been evolved,

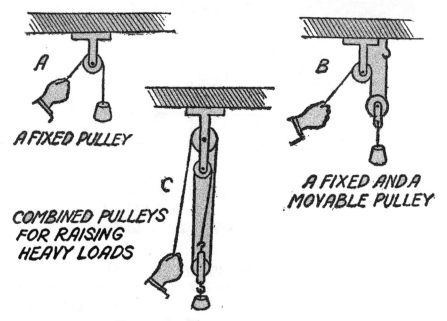

FIG. 34.—KINDS OF PULLEYS

with which any kind of a machine for any purpose can be built up.

To Find the Speed of a Shaft, Pulley or Fly-wheel.—To find the speed at which an engine runs, or a shaft, pulley or flywheel of a machine rotates, is a very easy matter if you have a *speed indicator* as shown at A in Fig. 35, to do it with.

This instrument is simply a worm gear, the spin-

dle of which has threads cut on it. These mesh with
the teeth of a gear to which an *indicator dial* is fixed.
To use the indicator, all you need to do is to set it at
0, note the time on your watch, and press the pointed
end of the spindle on the center of the end of the

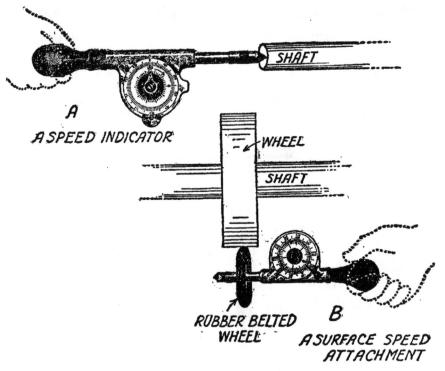

FIG. 35.—THE SPEED INDICATOR AND HOW IT IS USED

revolving shaft. In a minute read the dial, which
will give you the number of revolutions per minute.

To find the *surface speed,* that is, the number of
lineal feet per minute the *periphery,* or surface, of
the wheel is traveling, a *surface speed attachment* is
used; this consists of a rubber-banded wheel that can
be slipped over the spindle of the indicator, as shown
at B.

To use it, hold the wheel against the surface of the shaft or pulley a minute or so and then divide the number of revolutions, as shown on the dial, by 2; now since each revolution of the dial indicates 6 inches, and twice around equals a foot, the result will give you the number of feet the surface of the wheel is traveling.

How to Find the Size of a Pulley.—Very often, after you have found the speed of an engine, or a motor, you will want to know what size pulley you will need on a line of shafting belted to it to turn a given number of revolutions per minute.

This you can easily do by using the formula:

$$S = \frac{d\,R,}{r} \text{ or a little plainer,}$$

$$S = d \times R \div r$$

where S is the size of the pulley you must have on the shaft in diameter in inches, and is what you want to find,

d is the diameter in inches of the pulley on the flywheel shaft, which you know,

R is the R. P. M., namely, the number of revolutions per minute, of the flywheel which you get from the speed indicator, and

r is the number of R. P. M. you want to make the pulley on the shaft which is belted to the engine or motor revolve at.

Fig. 36 shows a belt driven pulley transmission system as described in the formula above, which will make it easy to understand.

How to Figure the Size of Belt Needed.—The next thing you will want to know is what sized belt you will need to transmit, or carry, a given horse power from the engine, or motor shaft, to the pulley on the countershaft, or machine.

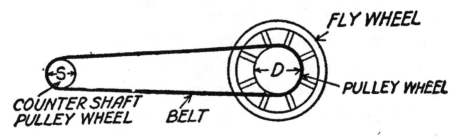

FIG. 36.—TRANSMISSION OF POWER BY PULLEYS AND BELTING

A belt larger than is needed will be an extra expense and result in a loss of power, while a belt that is too small will slip, break and behave badly in general. Hence, if you are to transmit power with economy, you must have a belt of the right width. You can roughly find about the right-sized belt from the formula:

$$W = \frac{H.P.}{7}$$

where W is the width of the belt traveling at 4,000 feet per minute (this is the most economical belt speed) and is what you want to find,

H.P. is the number of horse power to be transmitted and is known and

7 is a *constant*.[1]

How to Splice a Belt.—The three usual ways to join the ends of a belt [2] together are (1) to *cement* or *glue* them, when it is called an endless belt; (2) to *lace* them with rawhide; and (3) to fasten them with *metal* lacing.

To Make a Cement Splice.—A splice of this kind is shown at A in Fig. 37 and gives the least trouble

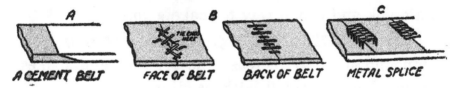

A CEMENT BELT   FACE OF BELT   BACK OF BELT   METAL SPLICE

FIG. 37.—KINDS OF BELT SPLICES

when done. However, it is hard to get the tension just right. Bevel off both ends of the belt with a block plane; then make a cement of 2 parts by measure of good liquid fish glue and 1 part by measure of Russian liquid isinglass; apply it to the beveled surfaces of the belt while it is hot, and then peg it with shoemaker's pegs $\frac{1}{2}$ inch apart.

To Lace a Belt.—Laces for belts are made of strips of rawhide and the width used varies with the size of the belt. Butt the ends together and punch two

[1] A *constant* is a fixed value that has been determined by experiment or calculation.

[2] For belting, lacing, etc., write the Page Belting Co., 152 Chambers Street, New York City.

rows of holes in each end, as shown at B. Begin at the center and lace it over to one edge, then back to the other edge, and then to the center again. Lace it so that the upper side of the lacing is parallel and crosses over on the under side.

Metal Belt Lacing.—This is a steel punching, as shown at C. To use it you only need to butt the ends together, set it evenly over both ends, and drive the sharp points through them. Turn the belt over, clinch the points and drive them into the belt. It is a quick and cheap way to make a joint.

A Good Belt Dressing.—A *belt dressing* is a compound used to increase the friction and makes the wheel pull the belt without slipping.

(1) Take 37 per cent of boiled linseed oil and mix it with 30 per cent of tallow; (2) mix 6 per cent of beeswax with 27 per cent of machine oil, all by measure. Heat 1 and 2 separately to 360 degrees *Fahrenheit* and then mix together.

Gears and Toothed Wheels.—In mechanics the word *gear* is used to mean two different things, namely, (1) a gear is a gear wheel, or cogwheel, as it is commonly called, that is, a wheel with teeth cut on its rim, or *periphery,* which can *mesh* with another toothed wheel or toothed *rack,* and (2) a gear is made up of a whole set of parts of some mechanical device, as the *steering gear* of an automobile. The only kind of gears we will talk about here are *gear wheels.*

There are five ordinary kinds of gears,[3] and these are (1) *spur* gears, (2) *internal* gears, (3) *miter* gears, (4) *bevel* gears and (5) *crown* gears.

Spur Gears.—A spur gear is a gear with teeth on its periphery. The three usual forms of this kind of gear are (1) the *spoked* gear, (2) the *webbed* gear

A-SPOKED B-WEBBED C-PLAIN

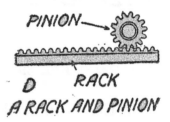

A RACK AND PINION

AN INTERNAL GEAR

FIG. 38.—KINDS OF SPUR GEARS

and (3) the *plain* gear, all of which are shown at A, B and C in Fig. 38. A rack is a flat strip of metal with teeth cut on one side of it, so that a spur gear which has the same sized teeth will mesh with it, as shown at D.

Internal Gears.—An internal gear is an inside, or

[3] For gears of all kinds, sprockets, etc., write the Chicago Model Works, 166 West Madison Street, Chicago, Ill., or Luther H. Wightman and Co., Boston, Mass.

ring gear, that is, it has teeth cut on the inside of its rim, as shown at E, so that a smaller spur gear can set in and mesh with it.

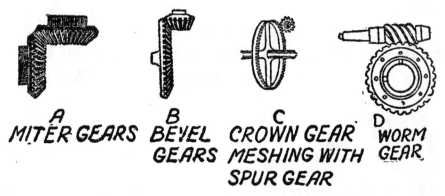

A
MITER GEARS

B
BEVEL
GEARS

C
CROWN GEAR
MESHING WITH
SPUR GEAR

D
WORM
GEAR

FIG. 39.—GEARS OF VARIOUS KINDS

Miter Gears.—Miter gears are gears of the same size set at right angles to each other with their teeth meshing together at 45 degrees, as shown at A in Fig. 39.

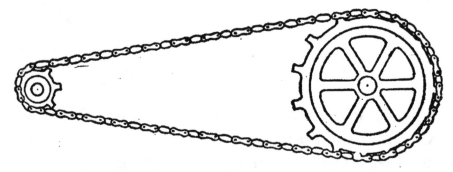

FIG. 40.—SPROCKET WHEELS AND CHAIN

Bevel Gears.—These are formed of two gears of different sizes set at right angles to each other and whose teeth mesh at any angle other than 45 degrees, as shown at B.

Crown Gears.—A crown gear, as shown at C, has its teeth cut on the edge of its face and, since it will mesh with a spur gear having teeth of the same size and pitch, the gears will very often serve just as well as bevel gears and have the advantage of running with spur gears of different diameters.

Worm Gear.—This is a screw working with a spiral gear as shown at D and is used in many machines for changing a high speed and small power into a slow speed and large power.

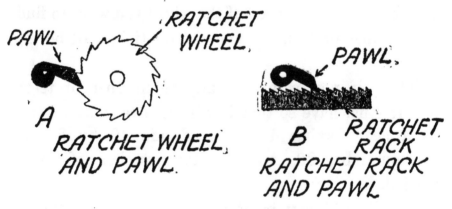

FIG. 41.—RATCHETS AND PAWLS

Sprocket Wheels.—Toothed wheels of this kind are used to transmit power by means of chains, as shown in Fig. 40, and, as there is no slippage, the *drive* is positive. For this reason they are often used in machines where there must be unity of action between the driving and driven shafts.

Ratchet Wheels, Ratchet Racks and Pawls.—A ratchet wheel has teeth cut on its periphery at a small angle so that it can be made to turn in one direction only and moved ahead a tooth at a time. This is done by means of a *pawl,* as shown at A in Fig. 41. Sometimes a *ratchet rack* and pawl are used to obtain an intermittent, horizontal motion of the former, as shown at B.

Figuring the Size of Gears.—(1) When you want to know *the number of teeth a gear must have to revolve at a given speed* when it is run by another gear the number of whose teeth you know, all you have to do is divide the number of teeth of the known gear by the rate of speed of the wheel you want to find the number of teeth on and the quotient will be the answer.

Thus, if you want to find the number of teeth a gear must have so that it will revolve twice as fast as a gear having 40 teeth, divide 40 by 2 and the quotient, 20, will be the number of teeth needed.

(2) Should you want to know *the speed that a gear will make with a gear whose number of teeth you know,* you only need to divide the number of teeth on the gear whose speed you want to find into the number of teeth on the gear whose rate of speed you know.

Thus, if a gear has 40 teeth and you want to know its speed when it meshes with another gear having 80 teeth which makes 20 revolutions per minute, divide 40 into 80. The answer, 2, will be the number of times it revolves to every complete revolution of the gear with 80 teeth; or 2 × 20, or 40, will be its number of revolutions per minute.

Friction and What It Does.—There is no such thing as a perfectly smooth surface. Even a sheet of highly polished glass has minute elevations and depressions on it, and, chiefly, because of these, if you lay one sheet of glass on another and slip it along it takes force to overcome the resistance, or *friction,* as it is called.

Now, while friction is a useful thing in our daily lives, since nails and screws would not hold and we could not walk and an automobile could not run without it, it is hard to contend with it in machinery, for it takes a lot of power to overcome it and this is wasted energy. The next best thing to do is to reduce the friction as much as possible, for this means to increase the efficiency of the machine.

How to Reduce Friction.—There are two kinds of friction, and these are (1) *sliding* friction and (2) *rolling* friction.

Where two surfaces slide on each other, one of them should be harder than the other to reduce the friction. Hence, steel shafts of machines are made to revolve in *bronze* or *babbitt* bearings. The friction between a rotating shaft and a fixed bearing is clearly sliding friction. The following are a couple of *anti-friction* alloys: [4]

[4] For bronze and babbitt write the Union Smelting and Refining Co., Avenue D and 14th Street, New York City.

TABLE

Name of Alloy	Copper	Antimony	Tin	Lead
Babbitt metal.............	4	7	89	
Bronze bearing..........	80		10	10

When a small quantity of phosphorus is added to the bronze alloy above, it forms what is called *standard phosphor* bronze bearing metal.

Rolling friction is very much less than sliding friction, and *roller bearings*,[5] as shown at A in Fig.

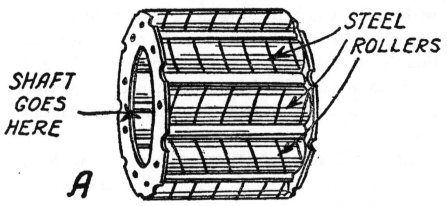

STEEL ROLLERS

SHAFT GOES HERE

A

FIG. 42A.—ROLLER BEARINGS

42, are, therefore, largely used in machinery. *Ball bearings*,[6] see B, offer still less resistance than roller bearings, because the surfaces in contact are not nearly as great; but there is some sliding friction

[5] For roller bearings write the Timken Roller Bearing Company, 1790 Broadway, New York City.

[6] For ball bearings write the Hess-Bright Company, 1914 Broadway, New York City, and to the New Departure Mfg. Co., Bristol, Conn.

even with ball bearings where the adjacent balls rub against each other or the *separators* which contain them.

The Use of Lubricants.—Wherever there is friction you can greatly reduce it by the use of a *lubricant,* but it must be a lubricant of the right kind.[7]

There are three factors to be considered in using a lubricant and these are (1) the pressure with which

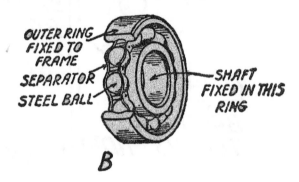

OUTER RING
FIXED TO
FRAME
SEPARATOR
STEEL BALL

SHAFT
FIXED IN THIS
RING

B

FIG. 42B.—A BALL BEARING

the surfaces slide against each other, (2) the speed that the surfaces are running at, and (3) how hot they get.

Vegetable, animal and mineral oils, soap, soapstone and graphite are used as lubricants, and each is good in its proper place. The following lubricants will serve as a key, but it must be remembered that there are many different grades of mineral oils.

(1) For watches, clocks and fine machinery, use

[7] For mineral lubricating oils write the Vacuum Oil Company, Rochester, N. Y., or the Platt and Washburn Co., 11 Broadway, New York City. For graphite lubricants write the Joseph Dixon Crucible Company, Jersey City, N. J.

olive oil that has been filtered, or add 1 ounce of kerosene to 2 ounces of sperm oil and filter.

(2) For machines that work at high speed and where the work is light, olive, rape, sperm or mineral oils can be used. The latter oil should have a

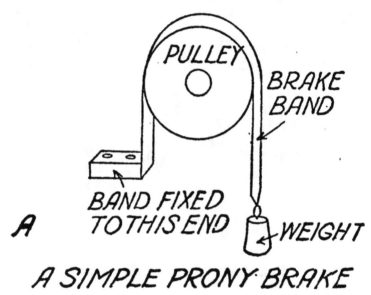

A SIMPLE PRONY BRAKE

FIG. 43A.—A DYNAMOMETER TO MEASURE HORSE POWER

specific gravity of 30.5 degrees *Baumé,* and a *flash point* of 360 degrees *Fahrenheit.*

(3) For ordinary machinery whale, neatsfoot, lard and heavy vegetable oils, vaseline and mineral oils are used. The latter should have a specific gravity of about 27 degrees *Baumé,* and a flash point of 400 to 450 degrees *Fahrenheit.*

(4) For cylinders of engines and other places where there are high temperatures, mineral oil having a specific gravity of 27 degrees *Baumé* and a flash test of 550 degrees *Fahrenheit* should be used.

These can be mixed with linseed or cotton seed oil or tallow.

(5) In slow speed and heavy pressure machines grease, soapstone or graphite can be used alone, or these can be mixed together.

(6) For wood use soap or graphite.

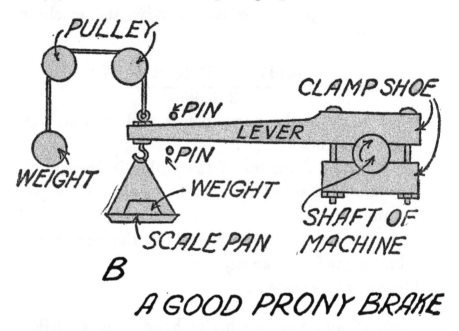

FIG. 43B.—DYNAMOMETER TO MEASURE THE HORSE POWER OF A MACHINE

How to Find the H. P. Needed to Drive a Machine.—When you have an engine or other source of power driving a machine, you can easily find the horse power needed to run the latter by means of a *dynamometer*.

A kind of dynamometer much used is called a *Prony brake.* In its simplest form it consists of a leather brake band which is slipped over the pulley of the machine, as shown at **A** in Fig. 43.   One end is fixed to a support and a weight is hung on its free end, which is just heavy enough to affect the speed of the pulley which you can tell by your speed indicator.

You can easily rig up a Prony brake and roughly find the horse power needed to drive the machine by the following formula:

$$\text{H.P.} = \frac{3.1416 \times D \times R \times W}{396,000}$$

where H.P. is the horse power and is what you want to find,

3.1416 is the diameter of the pulley in inches,

R is the number of revolutions of the pulley per minute,

W is the weight of the weight on the end of the brake band in pounds, and

396,000 is a constant.

Thus, if you put the brake band over a pulley 25 inches in diameter which is making 1056 revolutions per minute, and you find that a weight of 10 pounds hung on the band just slows down its speed, you cán find the horse power by substituting the figures for the formula above, thus:

$$\text{H.P.} = \frac{3.1416 \times 25 \times 1056 \times 10}{396,000} = 2$$

which means that 2 H.P. are needed to turn the pulley at that speed.

A better though more complicated form of Prony brake is shown at B, in which a pair of brake shoes are clamped around the shaft, and these absorb the power which turns it. This is the really practical type of dynamometer, but for your purposes the simple Prony brake will probably be accurate enough.

CHAPTER VI

PUTTING WIND AND WATER POWER TO WORK

The source of all the power we have that is available for useful work is the sun and the two chief natural powers due to it are (1) *wind power* and (2) *water power.*

WIND POWER

What Wind Power Is.—The *wind,* as we call it, is simply a current of air and this is caused by the sun heating some parts of it more than other parts. To equalize this difference of temperature the cold and heavy air flowing to the hot and lighter air sets it in motion when it develops power in virtue of its *weight* and *speed.*

Although the air is a yielding fluid, it acts just about like a solid body if it is moving swiftly enough or it is hit with something hard enough. Thus, when you fly a kite, the force of the wind drives the slanting kite up and out while you hold it in and down. But, if the string should break and the kite should keep the right slant, it would go on as long as the wind lasted.

The Parts of a Windmill.—There are seven chief parts to a real windmill,[1] of the kind that is used in the United States, and these are (1) the *tower,* (2) the *turntable,* (3) the *main shaft,* (4) the *wheel,*

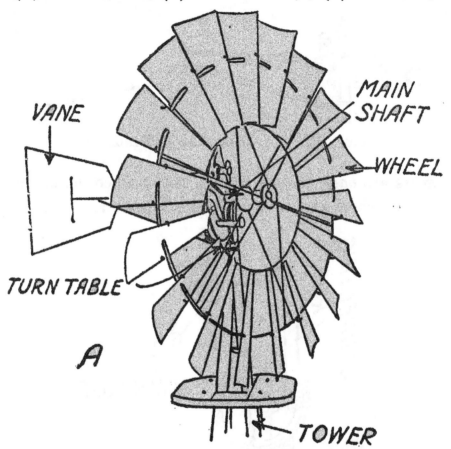

VANE

MAIN SHAFT

WHEEL

TURN TABLE

A

TOWER

FIG. 44A.—THE PARTS OF A STEEL WINDMILL

(5) the *gears,* (6) the *tailbone* and (7) the *vane,* or rudder, all of which are shown in Fig. 44A.

The turntable is mounted on top of the tower and connects the mill with it; the main shaft is fixed to

[1] For steel windmills, towers and pumps write to Woods and Co., 59 Park Place, New York City.

the wheel, which has radiating sails, as the *blades* are called. When the mill is used for pumping the

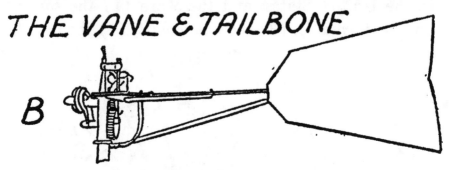

THE VANE & TAILBONE

B

FIG. 44B.—THE PARTS OF A WINDMILL

gears are *back geared,* that is, they reduce the speed of the wheel and so develop more power. A pump

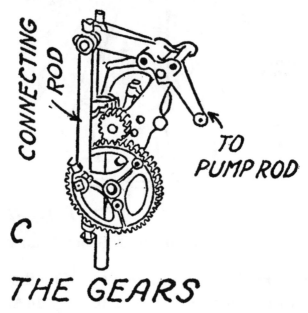

CONNECTING ROD

TO PUMP ROD

C

THE GEARS

FIG. 44C.—THE PARTS OF A WINDMILL

pole is connected to a crank on the small gear and to the pump below.

When the mill is used for running machinery, such as a feed cutter, sheller or wood saw, a beveled gear connects the main shaft with a small vertical shaft that runs down through a pipe, where another bevel gear changes the vertical rotary motion into a horizontal rotary motion. Thus not only is the power of the mill transmitted to the ground but the bevel gears step up the speed.

Sizes of Windmills for Pumping.—The following table shows the sizes of windmills required for wells of different depths:

TABLE

Size of Windmill	Depth of Well
6 foot wheel........................	25 foot
8 " "	50 "
10 " "	75 "
12, 14, 16 and 20 foot wheels........	For very deep wells

Sizes of Windmills for Machinery.—The following table gives the approximate horse power of mills working in winds of different speeds:

TABLE

Size of Wheel	Wind Velocities					
	10 miles per hour	15 miles per hour	20 miles per hour	25 miles per hour	30 miles per hour	35 miles per hour
12 foot wheel.....	.2	.67	1.6	3.12	5.4	8.5
16 " "	.36	1.21	2.9	5.5	8.5	15.3

The Height of Efficient Winds.—If your wind-mill is too low, the house, barn, trees, etc., will cut off the force of the wind and this will reduce its efficiency.

To get the best results, have the height of your windmill 15 or more feet above all wind obstacles. In any case, see to it that the tower is high enough so that the lightest wind blowing from any direction will have a clean sweep across the mill.

To Find the Height of Buildings, Trees, etc.— To know how high a tower you need to get the best results, stand a pole 10 feet high in the sunshine and measure the length of the shadow it casts. At the same time measure the length of the shadow cast by the highest building or tree nearest the place where you are going to set up your windmill.

Now divide the length of the shadow of the tree or house by the length of the shadow of the pole, and multiply the height of the pole by the quotient; or, to make a formula of it so that it will be easier,

$$H = \frac{L}{l}h$$

where H is the height of the house which you want
   to find,
L is the length of the shadow of the tree or house,
l is the length of the shadow of the pole, and
h is the height of the pole.

As an example, suppose that the shadow cast by the

10 foot pole is 8 feet and the shadow of the building is 32 feet, then

$$32 \div 8 = 4 \text{ and } 4 \times 10 = 40 \text{ feet}$$

which is the height of the house.

About Towers for Windmills.—Towers for windmills can be made of wood or of steel; the latter are the best, safest and cheapest in the long run. They are made with four posts, in three sizes, and in heights of from 20 to 80 feet.

The first size is for 6 to 10 foot mills, the second for 12 to 14 foot mills and the third for 16 and 20 foot mills. These towers are fitted with swinging pump-pole guides where the mill is to be used for pumping, and with shaft guides where it is to be used for running machinery.

WATER POWER

What Water Power Is.—Water power is developed by the flow or fall of water from a higher to a lower level.

The water is raised from a low level by evaporation, which is caused by the heat of the sun; the evaporated water then falls as rain on higher levels. Then it either flows or falls to a lower level and thus it is that the sun is really the source of water power.

By its weight, the force of its current and its *centrifugal force,* or a combination of them, it can be made to turn a wheel and so develop rotary power.

Kinds of Water Wheels.—There are several

99

kinds of water wheels, but the chief ones are (1) the *overshot* wheel, (2) the *breast* wheel, (3) the *undershot* wheel, (4) the *turbine* wheel and (5) the *jet* wheel.

The first three types of wheels are old fashioned and little used because they are very wasteful of

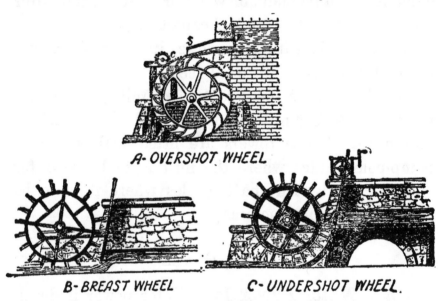

FIG. 45.—KINDS OF WATER WHEELS

the energy of the water and hence they must be large for the power they develop. They are shown in Fig. 45.

The Jet Water Wheel.—Where a small amount of water at a high pressure can be had, a jet wheel is the proper kind to use. This wheel has cups, or *buckets,* set around its rim, and the wheel, which is small for the horse power it develops as against the ordinary water wheel, is driven at a high speed by the force of the jet of water thrown on the buckets by a nozzle.

The Pelton water wheel [2] is the best known of this type. The wheel and nozzle can be mounted on a timber frame or encased in an iron housing. The water is discharged against the buckets by a

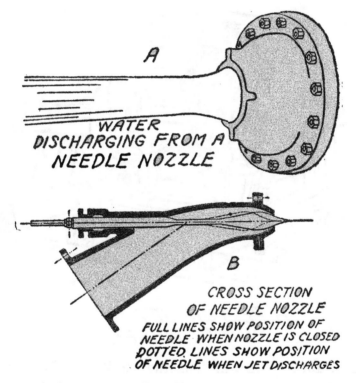

WATER DISCHARGING FROM A NEEDLE NOZZLE

CROSS SECTION OF NEEDLE NOZZLE
FULL LINES SHOW POSITION OF NEEDLE WHEN NOZZLE IS CLOSED
DOTTED LINES SHOW POSITION OF NEEDLE WHEN JET DISCHARGES

FIG. 46A & B.—THE JET TURBINE OR WATER WHEEL

specially designed *needle nozzle,* as shown at A and B in Fig. 46. This sets below the wheel, as shown at C.

The amount of water is controlled by moving the

[2] For data re the size and power of these wheels write to the Pelton Water Wheel Co., 90 West Street, New York.

needle in and out of the end of the nozzle either by
hand or by a governor geared to the main shaft.

The Water Turbine.—*Principle of the Turbine.*
—When water flows under pressure through a hose
pipe and out through a nozzle, it tends to straighten
out the hose. This is caused by the force of the

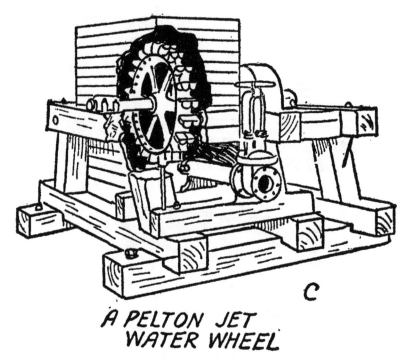

A PELTON JET
WATER WHEEL

FIG. 46C.—THE JET TURBINE OR WATER WHEEL

water rushing round the curved end of the nozzle,
that is, it whirls around and away from the center
because of its centrifugal force.

Now, in the turbine, the *centrifugal* force is produced by the water flowing through the curved fixed
guides when it strikes the guides, or buckets, of the
wheel, which are curved the other way, as the dia-

gram A in Fig. 47 clearly shows.   This kind of water wheel is the most efficient yet invented and it develops as high as 90 per cent of the total energy of the stream.

How the Turbine is Made and Works.—A vertical standard turbine[3] is shown at B and all the

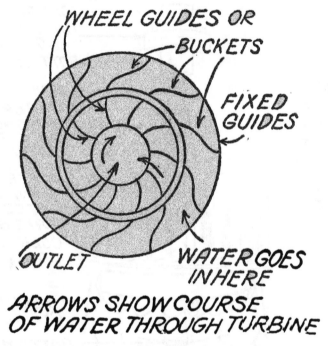

WHEEL GUIDES OR
BUCKETS
FIXED GUIDES
OUTLET
WATER GOES IN HERE
ARROWS SHOW COURSE OF WATER THROUGH TURBINE

FIG. 47A.—DIAGRAM OF HOW A WATER TURBINE WORKS

parts thereon are named.   The turbine sets on the floor of a *penstock,* or a *flume,* in an upright position and is entirely covered with water as shown at C.

The water from the penstock or flume is led to the turbine, which is set as low as possible so that the water flowing through it passes out of and into the

[3] For further information about turbines write to James Leffel and Co., Springfield, Ohio.

tailrace. In passing through the wheel, the water flows through the curved fixed guides when it is thrown on the buckets of the wheel in a direction that makes for the highest efficiency.

After the water has left the buckets the used water, or *tail water,* as it is called, flows out of the center

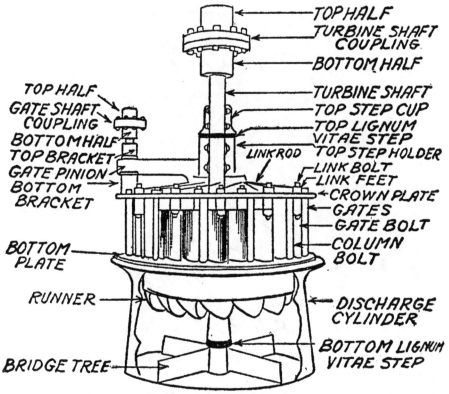

Fig. 47B.—A Standard Vertical Water Turbine

of the wheel, which is hollow, either directly into the tailrace or through a concrete or steel draft tube. The weight of the tail water in this tube produces a suction, which pulls the water from the penstock or flume into the wheel and makes it strike the buckets with greater force.

A vertical shaft is fixed to the turbine wheel and drives the machinery either by being connected direct to it, as in electric power plants, or by being geared to a driving pulley. Turbines are built in a large number of sizes and develop from 1 horse power with

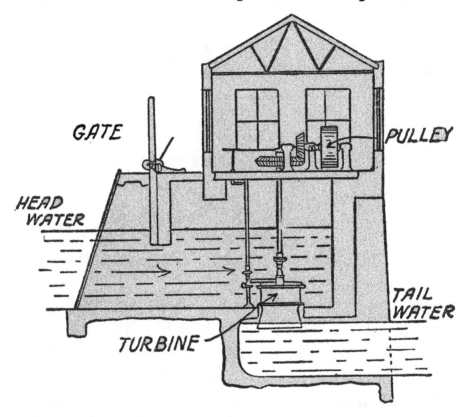

FIG. 47C.—THE WATER TURBINE AND HOW IT WORKS

a 3 foot head and a discharge of 252 cubic feet of water per minute, to 4000 horse power with a 50 foot head and a discharge of 51,100 cubic feet of water per minute.

The Hydraulic Ram.—This is a device used for raising water automatically to a considerable height

by means of a stream of water having a very small
fall. It has no revolving or moving parts except a
couple of valves, but it develops power in virtue of
the fact that, whenever the flow of a stream of water
is suddenly cut off, there is a corresponding increase
on the pressure of it.

A hydraulic ram consists of (1) the *body*, (2) an
air chamber, (3) a *sniff valve*, (4) a *check*, or *inlet*

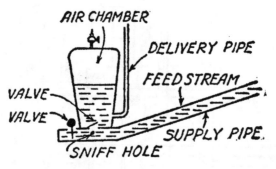

FIG. 48A.—CROSS SECTION OF A HYDRAULIC RAM

valve and (5) an *impetus valve*, all of which are
shown in the cross section at A in Fig. 48.

The hydraulic ram works like this: the water flows
down to the ram through a supply, or *drive pipe*, as
it is called, and out of the impetus valve at the end.
When the water gets a good start, the force of it
suddenly closes the valve and so cuts off the flowing
water.

This sudden stoppage sets up a high pressure in
the lower end of the pipe which forces the check
valve, set in between the drive pipe and the air chamber, to rise and open. Some of the water *rises* in
the air chamber and some of it is forced up through

the delivery pipe by the ramming blow of the water in the drive pipe.

As soon as the flow of water stops in the drive pipe, the impetus valve drops down and opens and the water again starts to flow in the drive pipe and out of the impetus valve; and then the cycle of operation begins all over again.

The space in the air chamber acts as a cushion for the water. This permits the check valve to open the moment the pressure is set up. The sniff valve is simply a small hole in the drive pipe, which sniffs in air for the air chamber, and it is sucked in when the recoil, or *kick,* resulting from the sudden rise of pressure, is set up. In this way water is constantly forced up in the delivery pipe.

A hydraulic ram is a cheap and satisfactory device for supplying water wherever a slight fall can be had. A small ram, having a capacity of from 60 to 100 gallons per hour and driving it to a height of 60 feet, can be bought for about $12.[4] It takes from 2 to 3 gallons per minute to operate the valves of this ram, which has a drive pipe of $\frac{3}{4}$ inch in diameter and a delivery pipe $\frac{1}{2}$ inch in diameter.

Larger hydraulic rams, taking from 2 to 700 gallons per minute to operate them, can be bought for from $50 to $850 each.[5] A Rife ram in action is shown at B.

[4] This ram is sold by the L. E. Knott Apparatus Co., Boston, Mass.

[5] These larger rams are made by the Rife Hydraulic Engine Mfg. Co., 90 West Street, New York City.

What "Head of Water" Means.—Before you install a water wheel, turbine or ram, you should first find (1) the *head of water in feet* that you are going to use, (2) the *quantity of water in cubic feet* that flows per minute, and from these two factors a simple calculation will give you (3) the *horse power of the*

INTAKE
TANK
SUPPLY
PIPE

DELIVERY

FIG. 48B.—THE HYDRAULIC RAM AT WORK

water supply, and then you will know what size water wheel or turbine you should use.

By *head of water* is meant the distance the water actually falls to operate the wheel or ram. Now, there are two kinds of heads of water, and these are (1) the *static,* or *surveyed head,* and (2) the *net, running* or *effective head.*

The *static,* or *surveyed head* as it is called, is simply the height of water in the penstock or where it flows into the flume or pipe measured to the lower level of the water wheel, turbine or hydraulic ram, or to the center line of the nozzle where a jet wheel is used. To measure the static, or surveyed head, use a carpenter's level and a yardstick, as shown in Fig. 49.

The *net, running* or *effective head* is the pressure of the water flowing in the penstock, flume or pipe.

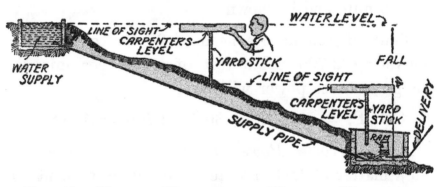

FIG. 49.—HOW TO MEASURE THE HEAD OF WATER OF YOUR SUPPLY

There are quite a number of factors which cause a loss of pressure from the static head, the chief one of which is friction. For a rough calculation, though, you can use the static head and let it go at that.

You can find the *quantity of water* flowing in a penstock, flume or pipe by catching and measuring the volume of water which flows out of them in 1 minute in cubic feet.

To Find the Horse Power of a Water Wheel.

—Finally, from the head and quantity you can easily

calculate the *gross horse power* of the **water wheel** by means of this formula:

$$G.H.P. = .00189 \times H \times Q$$

where G.H.P. is the gross horse power and is
   what you want to find,
.00189 is a constant,
H is the head in feet and which you have measured,
   and Q is the quantity of water in cubic feet per
   minute and is known.

Thus, if you have a head of 30 feet and a pipe delivering 2,700 cubic feet of water per minute, the gross horse power will be

$$G.H.P. = .00189 \times 30 \times 2,700 = 153.29$$

Actual Horse Power of the Water Wheel.— As a matter of fact, a water wheel is only about 80 per cent efficient and to find the *actual horse power* of the water wheel, you will have to multiply the gross horse power by .80. Then, in the preceding example, the *actual horse power* of the water wheel is

$$153.29 \times .80, \text{ or only } 122.63.$$

To Find the Amount of Water Delivered by a Ram.—You can find the amount of water delivered by a hydraulic ram from the following formula:

$$G = \frac{H \times Q \times 40}{D}$$

where G is the number of gallons delivered and is
   what you want to know,

H is the head in feet and which you know,

Q is the quantity of water in gallons (not cubic feet) per minute and which you know,

40 is a *constant,* and

D is the height you want the water delivered to.

Thus, if you have a head of 30 feet and a pipe delivering 2,700 gallons per minute and you want the ram to deliver this amount of water at a height of 84 feet, the amount of water delivered per hour will be

$$G = \frac{30 \times 700 \times 40}{84} = \frac{840,000}{84} = 10,000 \text{ gallons}$$

per hour.

CHAPTER VII

MAKING THE STEAM ENGINE WORK FOR YOU

Steam is the great prime power and it has done more to aid and abet civilization than all the other powers put together. To generate steam a boiler must be used, and to make the steam develop power an engine is necessary.

Now while a steam boiler and engine, or *power plant* as it is called, costs more to buy and to run than a windmill or a water wheel, a gas, gasoline or an oil engine, it is a far more certain source of power than any of these and it runs more smoothly and starts off with the full load the instant the steam is turned on.

About the Energy of Steam.—When 1 cubic inch of water is heated and changed into steam the latter will expand until it takes up nearly 1 cubic foot of space.

When water is heated to 212 degrees *Fahrenheit* it boils, and the more heat you apply to the water the more steam you will get and the hotter it will be.

Steam which can be seen is not real steam at all, but merely little drops of water that have been condensed by the cold air and carried up by the real steam, which is much hotter and quite invisible.

Now the heat of steam is of two kinds and these are (1) *kinetic* heat, that is, heat which makes the steam move, and this is what we call *sensible heat,* and (2) *potential heat,* that is, heat that is stored up in the steam, or *latent heat,* as it is called.

One of the curious things about energy of any kind is that it can be changed from energy of motion to energy at rest, and the other way about, with wonderful facility and quickness. Hence sensible heat can be changed into latent heat and *vice versa.* In an engine it does this in such a way that all the power there is in the steam is gotten out of it.

What Steam Pressure Is.—When water is heated to make steam the particles of water, or *molecules,* as they are called, are torn off from it and these are forced out in straight lines like miniature cannon balls. They keep on going until they hit other molecules or strike the sides of the vessel containing them.

This continual pounding away of the molecules of steam inside the boiler or the cylinder of an engine is so swift and hard that it sets up streams of force in every direction and this force and the extent of it is what is meant by the term *steam pressure.*

How Steam is Measured.—In this country steam pressure is measured in *pounds,* and this is done by connecting a *steam gauge* to the boiler near the top where the steam is hottest. The pressure of the steam acts on a mechanism that makes a needle swing over a dial, which is graduated to read in pounds.

Its action is just about the same as a butcher's scale when a piece of meat, or other commodity, is being weighed. The construction of the steam gauge will be explained presently.

How a Steam Boiler Is Made.—Different from steam heating boilers, those for running engines are built to develop and withstand *high pressures.*

There are two kinds of boilers in general use and these are (1) the *upright tubular* boiler and (2) the *horizontal tubular* boiler. Horizontal tubular boilers are of two kinds and these are (1) the *plain,* or *locomotive,* type and (2) the *return* type.

All of these boilers are the same in principle and are made up of three parts, namely, (1) the *boiler* proper, (2) the *fire box* and (3) the *smoke* box. Small boilers are nearly always of the upright kind and the larger boilers are generally of the horizontal kind. A boiler of either kind is a cylindrical shell formed of steel plates riveted together and having a *head* riveted to each end.

One large hole, or a number of small ones, are bored in each head and a single tube, called a *flue,* but more often a number of small tubes, called *fire tubes,* are fitted into them, as shown in Fig. 50. These tubes are made steam tight by *expanding* the ends of them, that is, spreading them out all round.

The fire box is an extension of and is riveted to the boiler shell and in it the grate is placed. The smoke box is either riveted to the other end of the boiler or else is made in the form of a hood to set

on it, while the smoke stack is bolted to the top of
the smoke box.

In the locomotive type of boiler the heat and smoke
from the furnace pass through the fire tubes in one
direction only, then out of the smoke box and through

FIG. 50.—A HORIZONTAL TUBULAR BOILER

the stack. In the return tube boiler the smoke box
sets on the same end as and over the fire box, so that
the heat and smoke pass through the tubes to the
front end and thence back again to the smoke box,
as shown in Fig. 51.

The Fittings of a Boiler.—Before a boiler can

be used to get up steam it must have a number of fittings. Chief among these are (1) the *water intake pipe*, (2) the *water pump*, (3) the *water gauge*, (4) the *steam delivery pipe*, (5) the *steam gauge cocks*, (6) the *steam gauge*, (7) the *safety valve* and (8) the *steam whistle*.

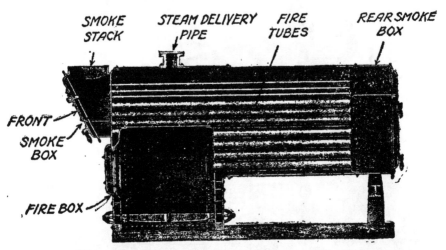

FIG. 51.—THE RETURN TUBULAR BOILER

(1) The water intake pipe connects the lower part of the boiler below the water line with a source of water. An ordinary *globe valve* is fitted to the intake pipe near the boiler and (2) a force pump is coupled to this and to the water supply to feed the water into the boiler against its *back pressure*.

(3) The water gauge is fitted to the shell of the boiler at the water line. It is formed of a long, upright glass tube set in two *wheel valves*, both of which connect with the boiler, as shown at **A** in Fig. 52. Since water seeks its own level, whatever the size, shape or position of the connecting vessel may

be, the level of the water in the glass gauge will be the same as that of the water in the boiler.

The water gauge is made so that the glass tube can be easily taken out and a new one put in without leaking. This is done by screwing a nut on each angle valve which has an opening in it large enough

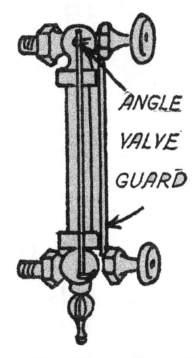

Fig. 52A.—The Water Gauge Complete

to take a rubber ring or washer. After the glass is slipped into place the nut is screwed up. This presses on the rubber ring and squeezes it until it fits tight against the glass tube as shown in Fig. 52B.

The best kind of tubes are called *Scotch glass* and these come in various sizes and lengths for different pressures. You can cut the tubes to fit by nicking them with a file, or, better, use a regular water gauge glass cutter.[1]

(4) The steam delivery pipe is screwed in the top of the boiler, if it is an upright one, or in the steam dome, if it is of the horizontal type, so that the hot-

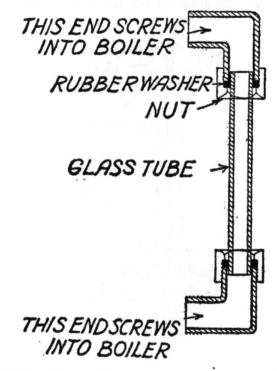

THIS END SCREWS INTO BOILER

RUBBER WASHER

NUT

GLASS TUBE

THIS END SCREWS INTO BOILER

FIG. 52B.—CROSS SECTIONS OF A WATER GAUGE

test steam, which has the most energy in it, will be delivered to the engine. A globe valve is fitted to this pipe near the boiler so that the steam can be cut off at this point if needs be.

(5) Three gauge cocks are fitted into the shell of

[1] These can be bought of Hammacher, Schlemmer and Co., Fourth Avenue and 13th Street, New York City.

the boiler just above the water line, and these are used to test the quality of the steam. Each one is fitted with a *stuffing box,* and they are shown on the right-hand side of the boiler in Fig. 50.

(6) To accurately measure the pressure of steam, a *Bourdon spring gauge,* see A, Fig. 52C, so called after its inventor, is used. It is made of a brass tube having a more or less flat cross section, which is bent into a ring, nearly, with its flat sides in and out, as shown at C.

One end of the tube is fixed to the frame of the gauge and the other end is open and is connected to the boiler through a bent pipe called a siphon. The other end is closed, and this end, which is free to move, is connected by a lever to a toothed segment which meshes with a pinion pivoted to the frame. A hand is fixed to the end of the spindle and this is turned back to its 0 position, when there is no steam pressure, by a spiral spring. When in use, the siphon is filled with water to keep the steam from directly reaching the gauge.

Now when the pressure of the steam is impressed on the flat tube by the siphon of water it tends to round it out; this makes the ring straighten out a trifle; in so doing it pulls on the lever, which moves the hand over the dial. The construction of the original Bourdon gauge is shown at B.

In attaching a steam gauge to a boiler, be sure that the siphon is filled with cold water. If the hand oscillates when the gauge is under pressure, close

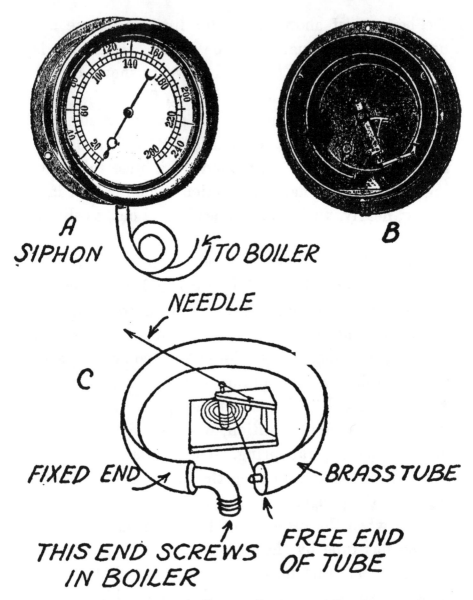

FIG. 52C.—A STEAM PRESSURE GAUGE

the cock a little, but not enough to reduce the pressure on the gauge. Always buy a gauge that is graduated to double the working pressure of the boiler, as this will insure accuracy. Fig. 52C.

(7) Every boiler must have a safety valve, so that the steam will blow off automatically before the pressure becomes dangerous. The safety valve for stationary engines is usually of the weight and lever type, as shown in Fig. 52D.

It consists of a valve in which a conical plug fits into a similarly shaped opening; this plug is held in its *seat* by a lever pivoted to the boiler at one end

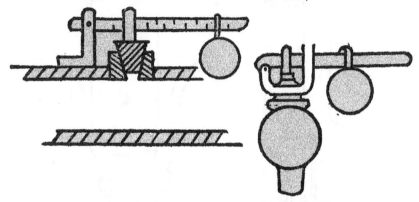

FIG. 52D.—HOW A SAFETY VALVE WORKS

and carrying a sliding weight on its free end. To make the valve blow off when a given pressure is reached, all you have to do is slide the weight along the graduated lever to the notch marked with the number of pounds you want.

A *bell whistle* is the kind that is blown by steam and is so called because the steam striking a cylindrical, open-mouthed tube makes it vibrate like a bell.

The whistle is formed of a heavy piece of tube closed at one end for the bell. This is fixed to a cup by means of a standard, and the cup in turn is fastened to the stem of a stopcock. Holes are drilled

in the top of the stem so that the steam can escape
in the cup when it strikes against the hollow side of
it and is forced up on the edge of the bell, which sets
the bell to vibrating. Then it gives forth a lusty
sound that everybody has heard. Its construction
is shown in Fig. 52E.

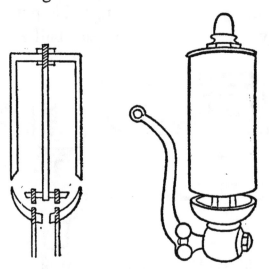

FIG. 52E.—HOW A STEAM WHISTLE IS MADE

Sizes of Steam Boilers.—A boiler should always
have twice the horse power of the engine it is to run.
Smaller boilers of $\frac{1}{4}$ *boiler horse power* to $1\frac{1}{2}$ B.H.P.,
in which gas, gasoline, kerosene, alcohol, wood or
coal can be burned, are made by the Lipp Electric
and Machine Co., Paterson, N. J. Larger boilers,
both upright and horizontal, can be bought of Done-
gan and Swift, 6 Murray Street, New York City.

How a Steam Engine is Made.—The steam engine is a machine for changing the energy of steam
into mechanical motion. Now, since steam is gen-

erated by heat and mechanical motion is power, what the steam engine really does is to change the heat into useful power.

But there are large energy losses from the time the fuel is burned in the fire box to the time the crankshaft of the engine is rotated. At the very best, not more than 20 per cent of the available energy that is in the fuel is changed into rotary power, and more often the efficiency is only 10 per cent, or even less.

Like steam boilers, there are two generic forms of steam engines. These are (1) the *upright engine* and (2) the *horizontal engine*. All ordinary engines, though, are made and work on the same principle, which I shall describe presently.

The Parts of an Engine.—For the purpose of explaining the steam engine, let's take one of the horizontal type, because its construction can be seen to better advantage than that of an upright engine.

There are sixteen chief parts to a steam engine, and these are (1) the *steam chest,* (2) the *slide valve* and its *stem,* (3) the *slide valve stem guide,* (4) the *eccentric rod,* (5) the *eccentric,* (6) the *cylinder,* (7) the *piston,* (8) the *piston rod,* (9) the *cross-head,* (10) the *cross-head guide,* (11) the *connecting rod,* (12) the *crankshaft,* (13) the *flywheel,* (14) the *pulley,* (15) the *pillow blocks,* and (16) the *bed.*

The steam chest is a box or chest through which the steam from the boiler passes into the cylinder. To make the steam flow first into one end of the cylinder and then into the other end a slide valve is used; this valve is a hollowed out metal block that covers alternately the *intake ports* of the chest which lead through ducts into the cylinder, and it also covers the *exhaust port* all of the time.

The slide valve stem is fixed to the slide valve and passes out of the steam chest through a *stuffing box,* that is, a chamber a little larger than the stem and in which *hemp* or other packing is stuffed to prevent the steam from leaking out when the stem slides forth and back.

The end of the slide valve stem slides through its guide; to the stem is pivoted the eccentric rod and the latter, in turn, carries the eccentric on the end of it. The eccentric is formed of a metal disk and this is mounted out of its center on the crankshaft. The disk has a groove in its rim and a collar or a strap is fitted into the groove, and this is connected to the eccentric rod.

The cylinder, as its name implies, is simply a cylinder with a solid head at the back and a front head with a hole in its center, over which is a stuffing box. The piston slides in the cylinder, and it is this element on which the steam acts. A piston rod is fixed to the piston and slips through the stuffing box on the head.

The other end of the piston rod is attached to the cross-head which is a metal block that slides in the cross-head guide. The connecting rod has a pair of bearings fitted to each end. One of these is pivoted to the pin of the cross-head block and the other to the pin of the crankshaft.

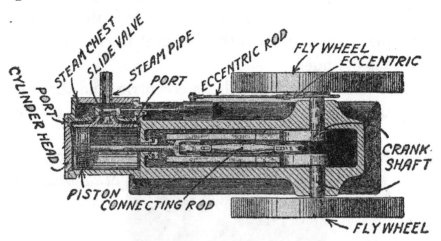

FIG. 53A.—TOP CROSS SECTION VIEW OF A STEAM ENGINE

The crankshaft revolves in a pair of bearings set in pillow blocks which support the crankshaft. The flywheel is keyed on one end and a pulley is keyed on

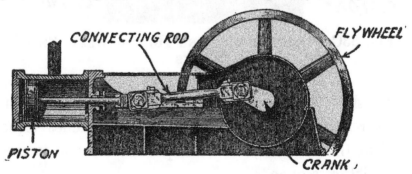

FIG. 53B.—SIDE CROSS SECTION VIEW OF A STEAM ENGINE

the other end of it. The cross section drawings of the top and side of an engine, as shown at **A** and **B** in Fig. 53, will make all parts of it clear.

How the Engine Works.—In the picture shown at C the steam chest is set above and away from the cylinder simply so that you can see to better advantage the ports and ducts that connect the steam chest with the cylinder.

The slide valve, through its eccentric, and the piston, through its connecting rod, are coupled to the

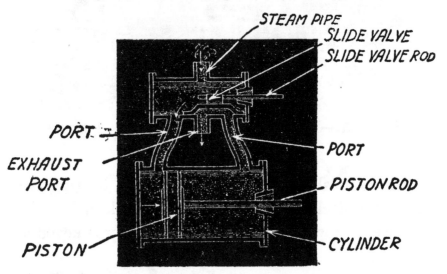

FIG. 53C.—DIAGRAM SHOWING HOW A STEAM ENGINE WORKS

crankshaft so that they move in opposite directions. The result is that, when the piston reaches either end of the cylinder, the inlet port at the end nearest the piston is open.

You will observe that the hollow in the slide valve is always over the exhaust port, and that it always covers the latter and one of the inlet ports at the same time.

Now the way the engine works is like this: When

the steam under pressure from the boiler passes into the steam chest, the slide valve is in one end of it and the piston is in the opposite end of the cylinder. Hence the port nearest to the piston is open and the steam flows through it into the cylinder and pushes the piston over to the other end.

When it reaches the port on this side it is open and the steam rushing into the cylinder forces the piston back again, which pushes the steam out of the other inlet port and thence, by means of the slide valve, out of the exhaust port into the open air.

Each forward movement of the piston pushes the crankshaft half way round and each backward movement pulls it the other half way round, thus making a complete *cycle*, or one revolution.

The Latent Heat of Steam.—The above is the simple mechanical action of the steam engine, but there is another factor which, though it cannot be seen, must be considered if the engine is to be an efficient one, and that is the *latent heat* in the steam.

Not only does the sensible heat of the steam produce pressure but the latent heat also; by this is meant that after the steam in the cylinder has been cut off by the slide valve, its latent heat, that is, the energy stored up in it, begins to change into energy of motion, and this makes the steam expand and keeps on pressing against the piston.

So to get all the power that is in the steam out of it, the length of the stroke of the slide valve is so adjusted that it cuts off the steam long before the pis-

ton has reached the end of its stroke, and the force
of the expansion of the steam is used to drive it the
rest of the way along.

What the Flywheel Does.—The flywheel accu-
mulates energy, which not only carries the crank
past its *dead centers,* that is, the ends of strokes
of the piston, but it also makes the engine run
smoothly.

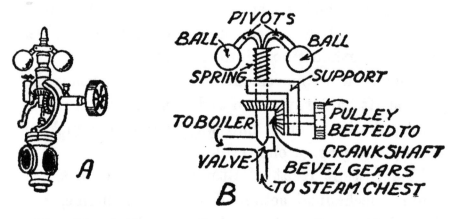

Fig. 54.—A Flyball Governor of a Steam Engine

How the Governor Acts.—A governor is used to
make the engine run at a constant speed. It does
this by regulating the flow of steam into the steam
chest.

The usual form of governor consists of an upright
spindle which is rotated by gears that are driven
by a belt from the crankshaft. Two levers are
pivoted so that their ends rest on top of the spindle,
and a ball is attached to each of the other ends. A
second pair of levers are pivoted to the first pair
and also a collar, which slides on the spindle, and this

in turn is attached to the valve of the delivery pipe.

When the engine runs too fast, the balls fly apart, which pulls the collar up and closes the valve. The instant the steam is cut off the engine slows down and the balls drop, thereby letting more steam into the steam chest and then the engine runs faster. A governor is shown in cross-section at B in Fig. 54 and just as it is at A.

Packing for Stuffing Boxes.—Packing [2] is used to prevent leakage around the piston and piston rod and the connecting rod. Formerly hemp was largely used for packing and the stuffing box was filled with it. To reduce friction and wear on the packing and rod, prepared packing was invented. This consists of flax, asbestos and rubber cemented together and lubricated with oil and graphite. It is quick and easy to put in and insures against leaks and blowouts.

How to Figure the Horse Power of a Boiler.—Since a boiler does not do mechanical work, the horse power of it cannot be calculated in the same way as in the case of an engine. It has been found by experiment that, when 34.5 pounds of water are changed into steam from and at a temperature of 212 degrees *Fahrenheit, 1 boiler horse power* is produced. A boiler horse power is the amount of steam power needed to run an engine of 1 horse power.

It has also been found that, to change 34.5 pounds

[2] For kinds and prices write to *The Crandall Packing Co.*, 136 Liberty Street, New York City, or to *The Johns-Manville Co.*, 41st Street and Madison Avenue, New York City.

of water into steam from and at 212 degrees, the boiler must have 10 square feet of *heating surface*. By heating surface is meant all of the boiler that the fire actually strikes plus the total area of all the fire tubes plus two-thirds the area of the smoke box.

Thus the heating surface required in a boiler to make enough steam to run an engine, the horse power of which you know, is

$$H.S. = H.P. \times 10$$

where H.S. is the heating surface which you want to find,

H.P. is the horse power of the engine which you know and

10 is the number of square feet of heating surface needed to generate 1 boiler horse power.

As an example, suppose you want to buy a boiler for an engine of 2 horse power. Then H.S. = 2 × 10, or 20 square feet of heating surface is needed to generate enough steam to run your engine at full load.

How to Figure the H.P. of Your Engine.— You can find, roughly, the horse power of a single cylinder steam engine by using this formula:

$$H.P. = \frac{P \times L \times A \times R}{33,000}$$

where H.P. is the horse power which you want to find,

P is the pressure of the steam on the piston and this you get from the steam gauge of the boiler,

L is the length of the piston stroke in feet,

A is the area of the piston head in square inches and is found by multiplying the radius of the piston squared by 3.14,

R is the number of revolutions of the crankshaft which you can find by a *speed indicator,* and

33,000 is the number of foot pounds which are equal to 1 horse power.

Suppose, now, you want to find the horse power of an engine whose cylinder is 4 inches in diameter; the stroke of the piston is 8 inches; the pressure on the piston is 40 pounds to the square inch and the crankshaft makes 300 revolutions per minute.

The area of the cylinder is then $3.14 \times R^2$ or $3.14 \times 2^2 = 3.14 \times 4 = 12.56$ square inches; the length of the stroke in feet is $\frac{8}{12}$ foot or $\frac{2}{3}$ foot or .66 foot.

Substituting these known values in the formula, you have

$$\frac{40 \times .66 \times 12.56 \times 300}{33,000}$$

$$\text{or } \frac{99,480}{33,000} = 3 \text{ horse power.}$$

CHAPTER VIII

USING HOT AIR, GAS, GASOLINE AND OIL ENGINES

Hot air, gas, gasoline and oil engines furnish sources of power that have many advantages for home use over windmills, water wheels and steam engines and, as each of the first named has its own peculiar qualities, these will be cited as we go along.

The Hot Air Engine.—While the hot air engine [1] is the most efficient of all heat engines, it is only used for pumping water, because of its small power compared to its size.

The chief advantage of the hot air engine lies in its absolute safety. Any boy or girl who can build a fire or light a gas jet or a kerosene burner can run the engine as well as a grown person. And, further, a little fuel is all that is needed to have a supply of water all the time.

How the Hot Air Engine Works.—The chief parts of a hot air engine are (1) the *displacement,* or *expansion,* cylinder, (2) the loose fitting *transfer piston,* (3) the *piston rod* and *connecting rod* for it,

[1] Hot air engines are sold by the Rider-Ericsson Co., 20 Murray Street, New York City.

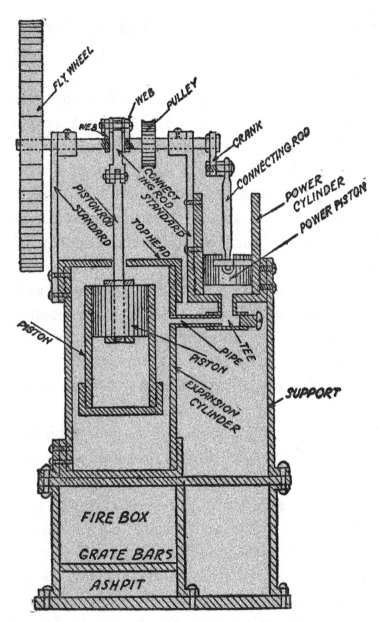

FIG. 55.—CROSS SECTION OF A HOT AIR ENGINE

(4) the *power cylinder*, (5) the *power piston*, (6) the *power piston rod* and *connecting rod*, (7) the *crankshaft*, with *pulley* and *flywheel*, (8) the *stand-*

ards, on which it is mounted, and (9) the *fire box,* all of which are shown in the cross section view in Fig. 55.

Now when a fire is built in the fire box it heats the bottom of the expansion cylinder and, on giving the flywheel a turn, the loose fitting transfer piston in it moves down. This forces the hot air in the bottom to go up and around it and into the top of the cylinder. On reaching the upper part the air is cooled by a water jacket around the cylinder in which water is flowing.

When the air is thus cooled it contracts, and, as the expansion and power cylinders are connected, the air contracts in the latter as well as in the former. The power piston is pushed down by the force of the air outside upon it, or the *atmospheric pressure* as it is called.

Since the transfer piston and the power piston are set at a *straight* angle, that is, an angle of 180 degrees, when the power piston is moving toward the bottom the transfer piston moves toward the top. This forces the cooled air back to the bottom of the expansion cylinder, where the fire heats it once more. When it is heated, the air expands, pushing the power piston up, and the cycle starts all over again.

Just bear in mind that the power is developed only in the power cylinder by the hot air expanding against the power piston first and, on cooling, by the atmospheric pressure outside of it.

How to Use a Hot Air Engine.—When burning

coal, a good draft is needed. A 5-inch stove pipe should be used for the smaller engines, and a 6-inch pipe for the large engines, and a damper must be put in the pipe in either case.

Chestnut size hard coal is the best fuel. This should be fed into the fire box in small quantities often in order to get an even heat and a steady speed. Kerosene and gasoline burners can be bought of the makers of hot air engines for burning these fuels.

The Gas Engine.—A gas engine is better than a steam engine and boiler in that its first cost is cheaper, it is smaller for the amount of power it gives, it does not need to be looked after so closely, and it is more economical to run.

Different from a steam engine, though, a gas engine must be run at its full working speed before it can be used to transmit its power to machinery. Otherwise it will *stall;* this is because it is the sudden force of the explosion of the gas that drives the piston to the end of the cylinder, while the heat in the steam makes it expand and develop power from the moment it enters the cylinders. Hence, a gas engine must be started up before the *load* is thrown on and this can be done either by shifting a belt or by using a *clutch* of some kind.

The Parts of a Gas Engine.—A gas engine is formed of the following principal parts, namely: (1) the *cylinder,* (2) the *piston* and its *connecting rod,* (3) the air and gas *inlet valve,* (4) the *exhaust valve,* (5) the *camshaft and cam,* (6) the *timing gears,* (7) the *crankshaft,* on which are the pulley and *flywheel,* and (8) the *igniter.*

In the type of gas engine in general use the cylinder is open at one end, as shown in Fig. 56. The piston is connected direct to the crankshaft by

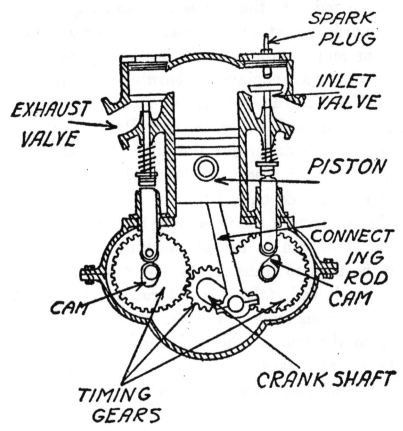

SPARK PLUG

INLET VALVE

EXHAUST VALVE

PISTON

CONNECTING ROD

CAM

CAM

CRANK SHAFT

TIMING GEARS

Fig. 56.—Cross Section of a Gas Engine

a connecting rod, and this does away with the piston rod, cross-head and cross-head guide.

The inlet valve is set in the closed end of the cylinder and works against a spiral spring. This lets the *fuel mixture,* as the gas and air which form the explosive charge is called, into the cylinder. A cam opens it at the right instant to admit the fuel mixture.

The exhaust valve is a valve in the head of the cylinder and this is opened at the right time to let out the burnt gases by the cam on the camshaft, which is geared to the crankshaft with a pair of bevel gears,

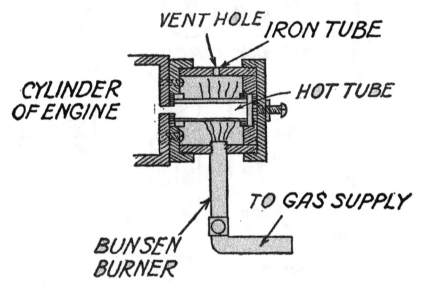

FIG. 57.—HOT TUBE IGNITER FOR A GAS ENGINE

or *timing gears,* as they are called. In automobile engines both the inlet and the exhaust valves are opened by cams on the camshaft.

The igniter which fires the fuel charge in the cylinder is set in the head of the engine. There are two kinds of igniters in general use and these are (1) the *hot tube igniter* and (2) *the electric spark system.*

The Hot Tube Igniter.—This is a very simple

137

kind and is still used on stationary gas engines. It is formed of a thin steel tube held in the middle of an iron shell by a cap on each end. The shell has a hole in it, and the steel tube is kept red-hot by a gas flame, as shown in Fig. 57. The igniter

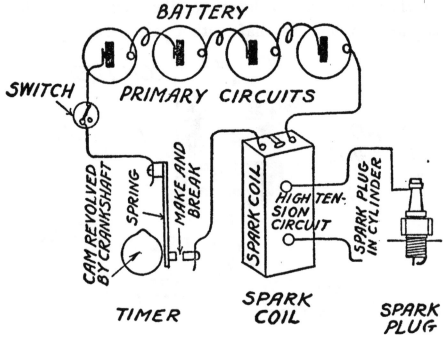

FIG. 58.—A BATTERY IGNITION SYSTEM

is screwed to the cylinder head over a hole in the latter, so that the fuel charge can be fired by it.

Electric Spark Systems.—There are two kinds of electric spark ignition systems, and these are (1) the *battery system* and (2) the *magneto system*.

The battery system consists of (a) a *dry* or *storage battery,* (b) a *spark coil,* (c) a *timer* and (d) a *spark plug.* These are connected up as shown in Fig. 58. The timer is a cam that is geared to the crankshaft and, when it rotates and makes contact with a spring, it closes the battery and spark coil circuit. A spark then takes place in the business end of the spark plug, which is screwed in the head of the cylinder.

The magneto system includes (a) a *high tension magneto,* (b) a *timer,* and (c) a *spark plug.* The magneto is a small dynamo electric machine [2] and

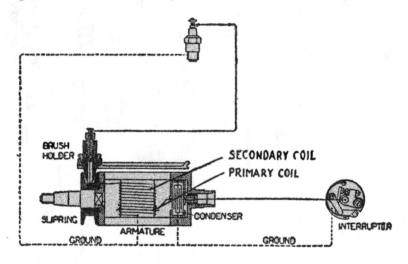

FIG. 59.—A MAGNETO IGNITION SYSTEM

induction coil combined, generating a high tension current. It is driven by a shaft geared to the crankshaft of the engine, as is also the timer. A diagram of the system is shown in Fig. 59.

How a Gas Engine Works.—A gas engine [3] works very differently from a steam engine, since in the first there is only one power stroke to every four

---

[2] The theory of the dynamo is explained in Chapter XI.
[3] Gas, gasoline and oil engines all work on the same general principle.

strokes of the piston, whereas in the second every stroke is a power stroke.

The diagrams shown in Fig. 60 represent a single cylinder gas engine, and each diagram shows a different stroke of the piston, also whether the valves are open or closed and what goes on in the cylinder.

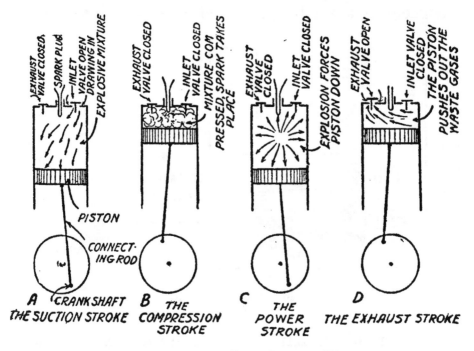

FIG. 60.—HOW A GAS ENGINE WORKS

To get a power stroke for every half turn of the crankshaft, the engine must have four cylinders whose pistons are connected with a single crankshaft. The valves are so timed that while one piston is making a suction stroke, the next is making a compression stroke, the following one a power stroke and the last an exhaust stroke. This gives the equivalent of a power stroke for every stroke of a single cylinder steam engine.

When a hot tube is used for the igniter the fuel charge is not fired until the piston reaches the end of its compression stroke, because the tube is not hot enough in itself to ignite the charge. But when any kind of a gas is compressed, heat is produced and this, together with the heat of the hot tube, increases the temperature to a point where it will explode the fuel charge. No mechanism is needed to make it explode at the right instant.

But, where an electric spark ignition system is used, a timer is necessary in order to make the spark at the end of the compression stroke.

How a Gasoline Engine Works.—While a gas engine burns ordinary city gas and a gasoline engine burns gasoline, the principle on which they work is the same.

The only difference between them lies in the fact that the former has an air and gas inlet valve or *mixing valve,* while the latter has a device called a *carburetor,* which breaks up the gasoline in a spray and mixes it with air when it is sucked into the cylinder.

The Parts and Action of the Carburetor.— A carburetor is made up of two chief parts and these are (1) the *gasoline supply control* and (2) the *spray making apparatus.* The gasoline supply is controlled, as you will see at **A** in Fig. 61, by a needle valve fitted to a float and as the gasoline fills the chamber the float rises and the needle valve cuts off the supply from the tank.

The spray making apparatus consists of a bent pipe connected with the float chamber and having a nozzle in its free end which is turned up.   Around the nozzle is placed a larger pipe as shown at B; one end

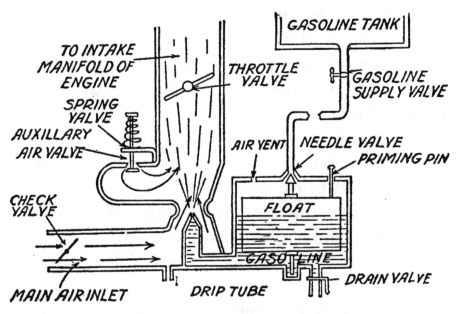

FIG. 61.—How a Carburetor Works

is open to the air and the other end connects with a *mixing chamber*. Now when air is drawn into this pipe it breaks up the gasoline flowing through the nozzle and this forms the *explosive mixture* which is drawn into the cylinder of the engine.

How an Oil Engine Works.—Oil engines are built on the same general lines as gasoline engines, hence they work on the same principle, but by using kerosene or crude oil for fuel these are safer and more economical to run.

The fuel oil is kept in a *supply tank,* which should be set below the level of the ground and outside of the building where the engine is placed, as shown in Fig. 62. The oil is pumped from the supply tank into a *fuel reservoir* fixed to the cylinder of the en-

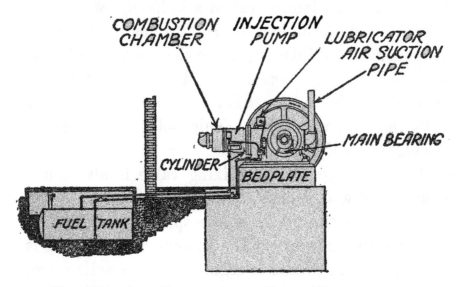

FIG. 62.—OIL ENGINE WITH TANK UNDERGROUND

gine, and the oil in it is kept at a constant level by an overflow pipe which carries the excess oil back to the supply tank.

From the fuel reservoir the oil flows into the *mixing valve* which breaks up the kerosene, or crude oil into a spray. The mixing valve is formed of a needle valve which sets in a nozzle; the small end of the nozzle is screwed into the head of the cylinder and the other and large end is connected to the fuel reservoir. When the engine is running a small quantity of the oil is drawn into the cylinder with the air on the suction stroke of the piston, while a needle valve regulates the amount of oil that is taken into the cylinder for each charge just as in an ordinary carburetor.

In the end of the air inlet pipe of the mixing valve is a little damper called a *butterfly valve*. By opening it more or less the right amount of air for the amount of oil used to make the proper fuel mixture for varying loads can be had. To do this a governor, called a *throttling governor*, is connected with the butterfly valve. This holds the speed of the engine steady.

The inlet pipe of the mixing valve is also sometimes fitted with a nozzle attached to a supply of water, which is thrown in a fine spray and drawn in with the fuel mixture. The amount of water that is taken into the cylinder is regulated by a needle valve.

The instant the water gets into the cylinder it is converted by the heat into steam. This acts as a cushion to break the violent force of the explosion and makes the operation of the engine more economical without reducing the power. The water must not be turned on until the engine has been running for some time, and it must be shut off a little while before the engine is stopped so that the cylinder will be left dry.

To start an oil engine, especially where crude oil is used, it is a good scheme to fill the reservoir with gasoline first. By the time this is used up the engine will be warm and work better.

Sizes and Power of Engines.—*The Hot Air Engine.*—The sizes of hot air engines are not based on the horse power which they develop, but on the vertical heights to which they can pump water. Thus an engine with a cylinder 5 inches in diameter will pump water to a height of 50 feet; 6 inches to 75 feet; 8 inches to 125 feet; and 10 inches to 160 feet. For data of hot air engines write to the Rider-Ericsson Engine Co., 20 Murray Street, New York City.

The Gas Engine.—Gas engines are built in all sizes, from 1 horse power on up to any horse power you want. Wherever there is a supply of natural or artificial gas, you have a source of power that is at once cheap and requires a minimum of attention. For data, floor space required, speed, weight and other data write to the Otto Gas Engine Works, 114 Liberty Street, New York City.

The Gasoline Engine.—A gasoline engine is not as economical to run as a gas engine but, where gas is not available, it is the next best kind of a prime mover. There are many makes of gasoline engines on the market, but to get a line on them write to the Otto Gas Engine Works and the Rider-Ericsson Engine Co., as above; Fairbanks, Morse and Co., 30 Church Street, New York City, and Sears, Roebuck & Co., Chicago.

The Oil Engine.—The smallest oil engine that I know of develops 2½ horse power, and from this little unit the sizes go on up to those large enough to run a sugar refinery or to supply power for a submarine.

An oil engine uses about half as much kerosene as the amount of gasoline used by a gasoline engine and, as kerosene costs about half as much as gasoline, it is obvious that it costs about a fourth as much to run it. Crude oils are even cheaper than kerosene, but it is better to run small engines on kerosene than on the heavier oils.[4]

For small oil engines write to Sears, Roebuck & Co., Chicago, Ill., and for the larger sizes get in touch with Fairbanks, Morse and Co.

How to Figure the Horse Power of a Gas, Gasoline or Oil Engine.—You can find about the number of horse power a four stroke cycle engine will give with this rule:

$$\text{H.P.} = \frac{D^2 \times N}{2.5}$$

where H.P. is the horse power you want to know,
D^2 is the bore or diameter of the cylinder squared,
N is the number of cylinders, and
2.5 is the coefficient and has been found accurate for a piston speed of 1,000 feet per minute.

[4] Before buying any kind of an internal combustion engine write to the *National Board of Fire Underwriters*, 76 William St., New York City, for a booklet called *Regulations for the Installation and Use of Internal Combustion Engine* which will be sent you free of charge.

USING OTHER HEAT ENGINES

Now suppose you have a 1 cylinder engine whose bore is 2¾ inches, then:

$$\text{H.P.} = \frac{2.75^2 \times 1}{2.5}$$

$$\text{or H.P.} = \frac{7.5 \times 1.}{2.5}$$

$$\text{or H.P.} = 3.$$

You will find the internal combustion engine more fully treated in my book, *Gas, Gasoline and Oil Engines*, published by D. Appleton and Co., New York.

CHAPTER IX

HOW TO HITCH UP POWER

Wherever you live you can easily have some kind of power and, having it, you can with a little scheming harness it up and make it pump water, wash clothes, saw wood and do a hundred and one other chores in and around the house and farm.

How to Use Wind Power.—While wind power is intermittent and variable, a good windmill properly fitted with transmission gears can be erected on top of your barn, either by using a four post mast or a steel tower, and running the vertical shaft down inside of it.

The lower end of the vertical shaft is geared to a horizontal shaft and this in turn has a pulley keyed to it. The drive is then braced securely to hold it in place, when it can be belted to whatever machine you want it to run.

When I say any machine, I mean a machine which does not require a constant speed, as, for instance, a feed cutter, corn sheller, circular saw for sawing wood, and the like.

About Changing Wind Power Into Electricity.—Many attempts have been made to generate elec-

148

tricity by using a windmill as a prime mover, but as the speed of the latter is so variable and the power is so uncertain, it is not to be recommended, especially since the oil engine is so cheap to install and to run.

How to Use Water Power.—If there is a stream of water on your place, you have a source of power that you can develop with very little trouble and at small initial expense. It will do all kinds of useful work without cost and with practically no attention after the plant is in operation.

All ordinary machinery can be belted directly to the pulley on the shaft of a water wheel, or you may have to use a pair of gears to speed up the drive pulley. If you should want, however, to transmit the power from the water to some distant point there are two ways open for you to do it and these are (1) by a *rope drive* and (2) by *electric transmission*.

To transmit power by a rope drive means simply that you use an endless hemp rope instead of a belt to connect the grooved pulley on the shaft of the water wheel to a similar grooved pulley at the distant place where you want to run the machinery. Where power is to be transmitted over short distances and light and heat are not needed at the other end, a rope drive is both cheap to install and to keep up.

The distance to which a rope drive will work satisfactorily, ranges anywhere from 10 to 175 feet, while with carrying pulleys the power can be transmitted to almost any distance.

149

Should you intend to install a rope transmission of any kind, write to the American Manufacturing Company, Noble and West Streets, Brooklyn, New York, for a copy of their *Blue Book of Rope Transmission* which they will send you gratis. In it you will find out everything that is known about transmission ropes and rope driving.

Where power is to be transmitted over considerable distances, the only feasible scheme is to belt or gear a dynamo to the water wheel and convert the energy of the head of water into current electricity. If the water wheel can be fitted with a governor to regulate the flow of water, the current can be used for lighting, as it comes direct from the dynamo.

But for a lighting system it is always good practice to hook up a storage battery to the dynamo and then pull the current from the storage battery. This arrangement not only gives a uniform current but, when the battery is charged, you can shut down the water wheel and dynamo, lock up the power house and leave it to the battery to deliver the current without fear of something going wrong.

How to Use Steam Power.—Steam is the ideal power for running all kinds of machinery in general and dynamos in particular, because it is steady, continuous and easy to regulate. Where gas, gasoline or oil can be used to fire the boiler, it takes but little work to keep a steam power plant going; but it isn't safe to let a boiler and a steam engine run alone for any length of time.

A dynamo can be belted to a steam engine and the flywheel is often used for the pulley so that the dynamo can be run at a high enough speed without using countershafting. If the engine is a high speed one, the *armature* of the dynamo, that is, the revolving element, can be connected direct to the crankshaft of the engine.[1]

A storage battery need not be used to take the current from the dynamo and then deliver it to the lighting and heating appliances, but the current can be used for lighting, and all other purposes you want to put it to, as it is generated by the dynamo, that is, where the engine is used only for running the dynamo; where other machinery is driven by the engine and there are variations in the load, or if you want a current when the engine is not running, or you want more current than the dynamo alone will give at certain hours of the day or night then, of course, you will have to install a storage battery.

Using Hot Air Power.—A hot air engine serves admirably for pumping water, running corn shellers and any kind of small machinery where a safe power is needed for short periods of time. It is not a good power, though, for driving a dynamo, even when a storage battery is used in connection with it.

How to Use Oil and Gasoline Power.—An engine burning kerosene is the cheapest and, next to the hot air engine, the safest kind of a portable prime

[1] See Chapter XI.

mover, not only in its first cost but in operation and in upkeep.

You can't beat it as a handy power producer on the farm, for it will do nearly everything but herd sheep and milk cows. But, if you want to generate electric power for lighting, you must use a storage battery between the dynamo and the lighting circuits.

A gasoline engine runs more smoothly than an oil engine but, unless you have a four cylinder engine, when you run the dynamo with it, if you do not run any other machinery at the same time you can get along without a storage battery.

In these days when so many second-hand cars have been relegated to the scrap heap, you can often pick up a car with a 20 or 30 horse power engine for as many or a fewer number of dollars.

Having it, you can leave the engine on the frame and mount the latter on a foundation of timbers, or you can loosen the bolts and take the engine off of the frame and set it on timbers or on a concrete foundation.

How to Use Your Automobile as a Power Plant.—In these days when every well-to-do farmer owns a motor car, it is easy to make it serve as a power plant for driving light machinery, or even a dynamo, in a pinch.

The drive is of the *friction type,* that is, the rear wheels of a motor car set on a pair of rollers and, when the engine is running, the friction between the rubber tires and the surface of the rollers causes the latter to revolve.

How to Make a Friction Drive.—To make the drive, cut off four pieces of 2 x 4 scantling and have each one 34 inches long. Bore a 1-inch hole through the thick side of two of the pieces 2 inches from each

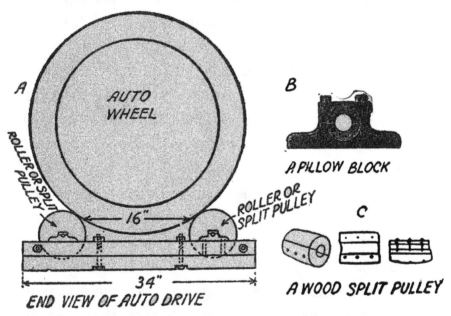

FIG. 63.—DETAILS OF AN AUTO POWER PLANT

end. Now get four *pillow blocks* and bolt one to each end of two of the sticks so that the center of the hole in it is 6 inches from the end, as shown at A in Fig. 63.

The pillow blocks are bearings made in two parts of cast iron. These are bolted together, as shown at B. A hole is drilled in the top part of the bearing so that it can be oiled. The hole for the shaft is 1 inch in diameter and the height from the center of it to the base is 1⅛ inches. You can buy them for about 75 cents apiece.[2]

After boring the holes for the bolts, round them out on one side so that the heads of the bolts will set in flush with the surface. This done, lay one of the sticks with the pillow block on it on one of the other sticks, bore a ¾-inch hole through both of them 11 inches from each end, and then bolt the sticks together. These form the ends of the drive.

Now get two pieces of iron rod 1 inch in diameter and 6 feet 2 inches long. Have a thread cut on each end of each one 6 inches down; screw a nut on each end and then slip on a washer. Put the ends of the rods through the holes in the ends in the sticks, slip a washer over each one and screw a nut on the end of each rod. This completes the frame.

The next thing is the rollers. You will either have to get these turned or else buy *split pulleys*,[3] that is, pulleys which are cut in two so that they can be bolted to a shaft, as shown at C. Get two lengths of steel shaft 1 inch in diameter. Have one of them 6 feet long and the other 6 feet 6 inches long. Then have a wood turner turn four hardwood rollers each of which is 6 inches in diameter and 12 inches long.

Next, bore a 1-inch hole down through the middle

[2] Luther H. Wightman, Milk Street, Boston, Mass., makes them.

[3] For further information and prices re wood split pulleys write the Dodge Sales and Engineering Co., 21 Murray Street, New York City.

of each roller. This must be done accurately, or else the roller will not run true. Drive these on the shafts far enough so that the ends of the latter can be set in the bearings of the pillow blocks, as shown at **A**, and then screw on the covers. On the projecting end of the long shaft, *key* or screw on a pulley to drive the machinery.

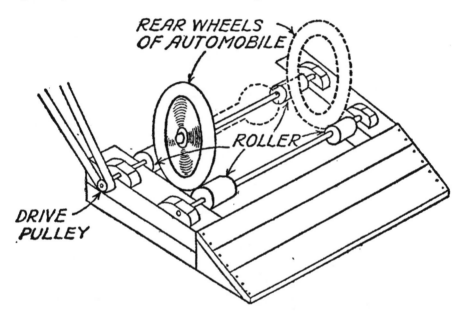

FIG. 64.—A MOTOR CAR POWER PLANT

Instead of having rollers turned, a better scheme is to buy four wood split pulleys 6 inches in diameter and having 8, 10 or 12 inch faces. These pulleys cost, respectively, about $3, $3.50 and $4 apiece. The construction of the pulleys is shown at C.

Finally, a runway must be made so that you can back the car onto the rollers and this is easily done by nailing a few boards to a couple of angle blocks.

Then your friction drive is ready to run as shown in Fig. 64.

To find the horse power of the engine, see Chapter VI; and to find the size and the speed of the drive pulley needed to run a machine at a given speed, see Chapter IV.

CHAPTER X

INSTALLING A HOME ICE-MAKING MACHINE

While it is easy to produce intense heat, it is quite another matter to make intense cold, especially on a small scale. Hence, ice is cut in the winter, stored in ice houses until summer and, when the thermometer is in the neighborhood of 100+ in the shade, it is delivered by the ice man at fabulous prices to the sweltering householder.

What Cold Is.—When we say a thing is *cold* we mean that it has a temperature which is lower than that of the *normal,* or standard, temperature, which is generally taken to be the temperature of the human body, namely, 98⅖ degrees *Fahrenheit.*

The standard of low temperature is the freezing point of water. This, as you found in Chapter IV, is 32 degrees on the *Fahrenheit* thermometer and 0 degree on the *centigrade* thermometer. But the freezing point of a substance does not by any means show that there is no more heat in it. The temperature at which a body really loses all of its heat is called the *absolute zero* and this is 273 degrees colder than the freezing, or 0 point, on the *centigrade* scale.

How Cold is Produced.—There is only one way by which cold can be produced and this is by *evaporation;* to do this a liquid must be used or if a gas is used it must be *condensed* into a liquid first, and in all ice-making machines both of these principles are combined and used.

Cooling by Evaporation.—In physics *evaporation* means that a vapor is formed on and given off by the exposed surface of water, or any other liquid, which has a temperature below the boiling point. In countries where the heat is intense, drinking water is kept cool by putting it in unglazed earthen jars, which are porous, and set in the shade where the wind will blow on them.

As the water seeps through the pores of the jar and reaches the surface, the wind evaporates it and this rapid evaporation keeps the water cool. An experiment to illustrate cooling by evaporation is to put a few drops of alcohol, or, better, ether, in the open palm of your hand, when it will evaporate very fast, and you will feel it get quite cold.

What Condensation Is.—In physics *condensation* means that a gas or vapor is changed to a liquid. Now, there are two ways a gas or vapor can be *liquefied* and these are (1) by cooling it and (2) by compressing it, and both of these processes are used in ice-making machines.

For experimental purposes and making ice cream, a freezing mixture can be made by mixing 1 part of salt with 3 parts of cracked ice, and this will produce a temperature lower than that of the freezing point of water. Again, if 3 parts of calcium chloride, which is a salt, are mixed with 2 parts of cracked ice, a still lower temperature can be had and one that is cold enough to easily freeze mercury.[1]

To compress a gas, or a vapor, until it liquefies, all that is needed is to draw it into the cylinder of a pump and push a piston against it. In ice-making machines the gas is cooled by cold water flowing in a coil of pipe, around which the gas circulates. It is then liquefied by compression in a pump.

About Ice-Making Machines.—Ice-making machines in general use today are worked with two kinds of chemicals for the refrigerants and these are (1) *ammonia gas* and (2) *sulphur dioxide gas.*

Ammonia Refrigerating Machines.—Machines for making ice on a large scale use *ammonia,* or *ammonia gas,* as it is called. This must not be confounded with the so-called *liquid ammonia* sold in stores, which is merely water that has absorbed a lot of ammonia gas and is really *ammonia water.*

Ammonia is a colorless, transparent gas. It is easily made into the liquid form when it is chilled and pressure is applied to it. When the pressure is removed from the liquefied ammonia it soon passes back to its gaseous state by evaporation. In so doing it absorbs heat and hence cools the surrounding air or water. These properties of it are taken advantage of in the artificial manufacture of ice. A

[1] Mercury freezes at —39.5° Fahrenheit.

cross section of an ammonia ice making machine is shown in Fig. 65.

Sulphur Dioxide Refrigerating Machines.—The only ice machine that is small, safe and economical

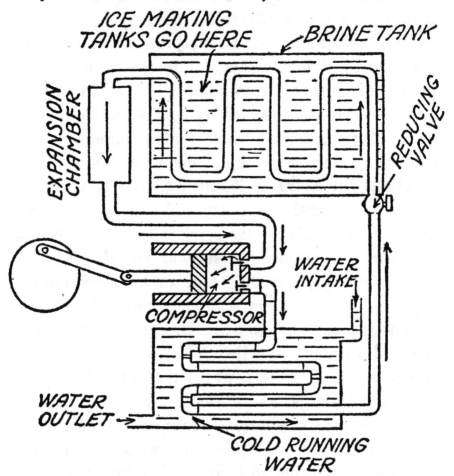

FIG. 65.—HOW AN AMMONIA ICE-MAKING PLANT WORKS

enough for home purposes that I know of is the one invented a dozen or fifteen years ago by Audiffren, a French physicist, and in which *sulphur dioxide* is used as the refrigerant.

Sulphur dioxide is a gas that liquefies much easier than ammonia gas, in fact all that is needed to liquefy it is to set the vessel containing it in a freezing mixture made of ice and salt, as previously described. Sulphur dioxide also liquefies at a much lower pressure than ammonia and has a much lower working pressure.

Different from the ammonia ice-making machine in which there is a leakage of the *refrigerant,* as the ammonia is called, through the stuffing boxes and pipe joints, the Audiffren sulphur dioxide machine has its refrigerant *hermetically* sealed, that is, sealed airtight, in the dumbbell which forms the rotating part of the machine.

The machine,[2] which is shown in cross section in Fig. 66, consists of a shaft with a pulley on one end, a *refrigerator drum,* or hollow shell, on the other end and a *compressor drum,* or hollow shell, set on the shaft between them. This revolving element, or *dumbbell,* so called from its shape, rests on two bearings, one on each side of the middle drum.

The *compressor,* as the pump is called, hangs on the shaft and it is held by a heavy lead weight, so that it always keeps an upright position. The piston which works in the cylinder of the pump is moved to and fro by means of a connecting rod fixed to an eccentric on the shaft.

Above the cylinder and the shaft is a reservoir for

[2] This ice machine is sold by The Johns-Manville Co., 41st St. and Madison Ave., New York City.

the liquid sulphur dioxide; this connects with the refrigerator through a float valve and pipe in the hollow shaft. The float valve automatically supplies the correct amount of refrigerant to the refrigerator drum through the pipe.

The refrigerator drum is fixed to the end of the

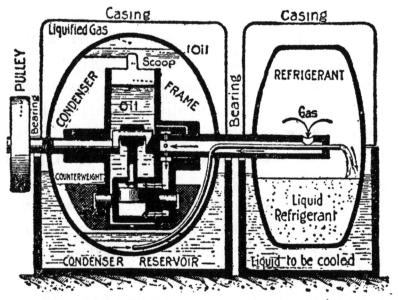

FIG. 66.—A SULPHUR DIOXIDE ICE-MAKING MACHINE

hollow shaft in which a small opening is left and the liquid sulphur dioxide flows from the compressor drum through the shaft and out of the hole in it in a spray into the refrigerator drum.

The latter revolves inside a tank of brine and when the evaporating sulphur dioxide has absorbed the heat of it the gas passes back through the space between the pipe and the hollow shaft to the compressor drum, where it is again compressed and liquefied.

The tank filled with brine is connected to a coil of pipe in the ice-making tank, or in a refrigerator or both as shown in Fig. 67. The brine that the refrigerator drum sets in is forced through the coils of pipe in the ice-making tank and refrigerator by

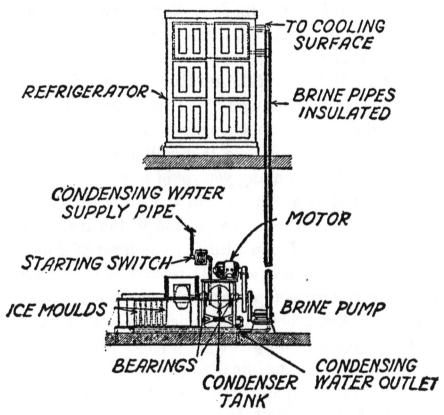

FIG. 67.—A COMPLETE ICE-MAKING PLANT

a small centrifugal pump called a *brine circulating pump.* This and the ice-making machine are driven by an electric motor, or some other source of power.

How to Insulate the Brine Mains.—To get the best results from an ice-making machine the outside pipes carrying the brine, or *brine mains,* as they are

163

called, must be well *insulated,* that is, covered, to prevent them from absorbing heat. The better they are insulated the smaller will be the expense of making the ice. The best insulation for brine mains is cork. This can be bought of the Armstrong Cork Company, 50 Church St., New York City, or of the Johns-Manville Company, 41st St. and Madison Ave., New York City.

How to Build a Refrigerator.—The refrigerator must also be thoroughly well insulated. If you will build one of the following materials in the order named, you will have one that will keep out heat as well or better than any you could buy.

The materials are named in the order in which they are built up from the outside to the inside of the refrigerator. Begin with (1) one layer of $\frac{7}{8}$-inch boards for the outside and (2) cover this on the inside with waterproof paper; (3) put on a layer of pure sheet cork 2 or 3 inches thick; (4) on this put another layer of waterproof paper; (5) then a layer of $\frac{7}{8}$-inch boards, and, finally, (6) line it with $\frac{7}{16}$-inch thick opaque glass, or thin sheet enameled steel.

Some Facts About Ice Making.—The following facts are interesting in connection with the making of ice. (1) Water that has been distilled will freeze clear, but it is not at all necessary to use distilled water to get nearly pure ice.

(2) When raw water is frozen it tends to force the impurities in it to the center. The slower the water is frozen the clearer the ice will be. (3) If

164

the water is stirred or otherwise agitated while it is freezing, the quicker and more surely will the impure matter be forced to the center of the cakes when it can be removed. Agitation helps to form clear, solid cakes of ice.

(4) The rate at which ice freezes decreases directly with the thickness of that which is already frozen. This being true, it follows that the time it takes to freeze a cake of ice increases in proportion to the square of the thickness of that to be frozen.

(5) To make raw water ice, fill the cans with the water and agitate it until it is partly frozen. Then draw off the remaining water. This will carry off most of the impurities that have been frozen out and into it. Fill the can with fresh water and agitate it while it is freezing as before.

What It Costs to Make Ice.—To find the difference in the cost of natural ice and mechanically made ice, you must include in the former (1) the cost of harvesting, which means the labor of cutting and storing it; (2) the melting and other wastage of it; and (3) the amount left over at the end of the season.

The cost to make ice with an ice machine varies within wide limits, too, but in any case it is based on the cost of (1) coal, (2) labor, (3) the refrigerant used, (4) water and (5) the power that is used, loss of oil, etc.

CHAPTER XI

ELECTRICITY IN THE HOME AND ON THE FARM

As you have seen from what has gone before, you can have power at very little expense wherever you live and, having it, you can convert it into electricity without the slightest trouble.

Now, while electricity is a *secondary power,* that is, it must first be generated by some other power such as water, steam or gas, you can do with it that which you cannot do with any of the others, that is, use it for light, heat and power at one and the same time.

What to Know About Electricity.—It is easy to understand how a current of electricity acts and works if you know just three things about it, and these are (1) that it has quantity, or *current strength,* as it is called, (2) that it has pressure, or *electromotive force,* as it is termed, to drive the current along, and (3) that the wire in which the current is flowing has *resistance,* that is, it opposes the flow of the current.

From this you will observe that an electric current behaves very like a current of water flowing through a pipe, hence, when you want to know how

166

the former would act under certain conditions, just consider what the latter would do and you will come pretty close to the right solution of the problem. You must be careful, though, not to carry this *hydraulic analogue* too far.

Current Strength and the Ampere.—When the *poles* of a battery, or a dynamo, are connected with a wire, or *circuit* as it is called, a current flows from the positive, or + side, to the negative, or — side.

Now the quantity of electricity, or *current strength,* or just *current* for short, as it is called, flowing in a wire or *circuit* depends on the *pressure,* or *electromotive force,* that is driving the current along the wire, and the *resistance* of the latter.

The greater the pressure, the larger the current that can be forced to flow through the wire; on the other hand, the higher the resistance of the wire, the smaller the current that can be forced through it.

To measure the amount of current that is flowing through a wire, or circuit, a unit called the *ampere* is used. 1 ampere is the amount of current that 1 *volt* of electromotive force will drive through a wire having a resistance of 1 *ohm.* The amount of current is measured by an instrument called an *ammeter.*

Electromotive Force and the Volt.—The pressure that forces electricity along a wire, or electromotive force, is measured by a unit called the *volt.* A volt is the electromotive force needed to drive 1 ampere through a wire having a resistance of 1 ohm.

The pressure or electromotive force is often called the *voltage* and it is measured with an instrument called a *voltmeter*. A dry cell gives a pressure of about 2 volts, and lamps, heating apparatus and motors for home electric plants are built to work with a pressure of 32 volts. Ordinary direct' current power plants generate current at 110 volts.

Resistance and the Ohm.—A wire of whatever size always resists the flow of a current through it. The resistance depends on the kind of metal the wire is made of, its diameter and its length.

The unit of resistance is the *ohm*. 1 ohm is the resistance of a circuit which requires a pressure of 1 volt to send a current of 1 ampere through it. An ordinary telegraph wire 400 feet long has a resistance of about 1 ohm. Resistance is measured with a *resistance box,* but it is easy to figure it if you know the current and voltage.

The Relation Between Current, Pressure and Resistance.—From the above you will see that there is a definite relation between current, pressure and resistance. This being true, it is obvious that, if you know the value of any two of them, you can easily figure out the value of the remaining one.

To do this just remember these three rules:

(1) That volts ÷ ohms = amperes;

(2) That amperes × resistance = volts; and

(3) That volts ÷ amperes = resistance.

With these fundamental laws in mind, you are ready now to get acquainted with the power plant and subsidiary apparatus for generating and using electric current.

What an Electric Installation Consists of.—There are four chief parts to an electric power plant and these are (1) the *prime mover,* or motive power; (2) the *dynamo,* which generates the current; (3) the *storage battery;* and (4) the *switchboard.*

The installation further consists of (5) the *transmission lines;* (6) the *service wires;* and (7) the *devices* that use the current. All the various powers that can be used for running dynamos have been described in the foregoing chapters and the methods by which these prime movers can be used to drive the dynamos have also been described.

The Dynamo Electric Machine.—There are two kinds of electric current used for lighting, heating and power and these are (1) *direct* current and (2) *alternating* current. While both of these can be used equally well for lighting and heating, direct current is better for running motors, and the cost of direct current motors is less than for alternating current motors.

Oppositely alternating current can be transmitted farther over smaller wires with less loss of power than direct current. But for all ordinary work direct current is the most satisfactory. Hence, for your power plant you should install a direct current machine, or *dynamo,* as it is called.

How a Dynamo Is Made.—A dynamo is a very simple machine and consists of two chief parts and these are (1) the *armature,* or revolving element in which the currents are set up, and (2) the *field mag-nets* between whose poles the armature rotates.

The armature is formed of a *core* of very soft iron and lengthwise on this a large number of turns of insulated copper wire are wound; the turns of wire are divided into *coils,* and the ends of each coil are connected to the opposite *segments* of a *commutator.*

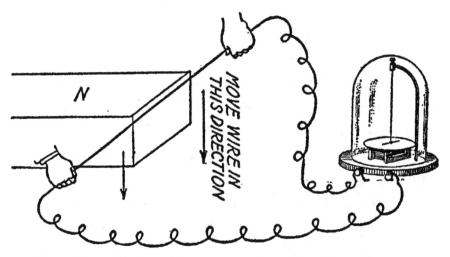

FIG. 68.—How a Current is Set Up in a Moving Wire

The commutator is made up of a number of copper segments, or bars, separated by strips of mica to insulate them from each other. Together they form a ring, and this is fixed to the shaft that carries the armature. As the currents are set up in the coils of the armature, they flow to the commutator bars, where they are taken off by a pair of soft carbon brushes which press on each side of the commutator. The field magnets are also made of very soft iron and these are wound with insulated copper wire.

How a Dynamo Generates Current.—A simple way to show how a dynamo generates a current is to connect the ends of a copper wire with a *galvanometer* and move the wire quickly across the pole of a magnet, as shown at **A** in Fig. 68.

The instant you do this the needle of the galvanom-

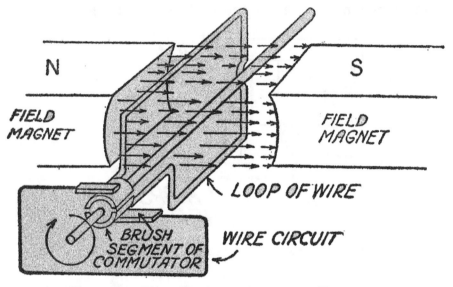

FIG. 69.—THE PRINCIPLE OF THE DYNAMO

eter will swing and this shows that a current is flowing in the circuit; further, this experiment shows that whenever a wire cuts the *magnetic lines of force,* the latter are changed into an *electric current* which is set up in the wire.

To make the wire cut the lines of magnetic force, form it into a loop, as shown in Fig. 69, and fix it to a spindle with a crank. When you turn the crank in the direction of the arrow, currents will be constantly set up in the loop of wire and will flow around it, first in one direction and then in the other, for every time the loop moves from one pole to the other the current set up in it changes its direction. Hence, there will be two alternations of the current for every revolution of the loop of wire.

By winding the wire on a cylinder of soft iron, the strength of the magnetic lines of force will be

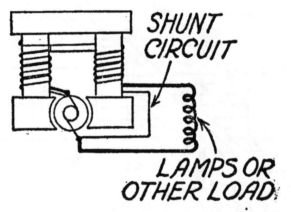

FIG. 70.—HOW A DYNAMO IS WOUND

greater, for magnetism flows through iron easier than through air and this, of course, increases the strength of the current set up in the wire.

To make the currents that are set up in the coils of the armature flow in one direction, the ends of the coils are connected with the segments of the commutator. For every coil on the armature, which is made up of a large number of turns of fine wire, there must be a pair of separate and oppositely set segments in the commutator.

A small part of the current taken off by the brushes

from the commutator flows back through the coils of the field magnets and so keeps them magnetized. In this way the magnetic lines of the fields are changed into electric currents by the armature, which generates enough additional current to light lamps, heat sadirons, wash clothes and do other useful work.

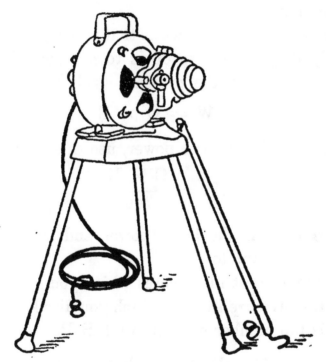

FIG. 71.—A PORTABLE ELECTRIC MOTOR

Dynamos are wound in different ways but the kind you want for your lighting plant is a *compound wound dynamo* as shown in Fig 70.

The Electric Motor.—Away back in Centennial days, that is, in 1876, some one found that if a current was passed through a dynamo it would run as a motor and develop power; Fig. 71 shows a portable motor capable of doing all kinds of work wherever you want it done. To find the horse power an electric motor is using or a dynamo is delivering in current, or a lamp or any other piece of electrical apparatus takes, you should know first that the *unit of electric power* is the *watt* and that there are 746 watts in 1 horse power.

To find the number of watts that is being generated or used, all you have to do is to multiply the current (amperes) by the pressure (volts)

$$\text{or } W = C \times E$$

Then to find the horse power, use this formula:

$$\text{H.P.} = \frac{C \times E}{746}$$

where H.P. is the horse power and is what you
want to find,
C is the current in amperes which you know,
E is the pressure in volts which you also know and
746 is the number of watts in 1 H.P.

Thus, if a 30 volt motor takes 6 amperes to run it, substitute these values for those in the formula to find the horse power developed thus:

$$\text{H.P.} = \frac{6 \times 30}{746}$$

$$\text{or H.P.} = \frac{180}{746}$$

or H.P. = .23 or very nearly ¼ horse power.

How a Storage Battery Is Made.—When two lead plates are set in a jar of dilute sulphuric acid, they form the simplest kind of a *storage battery.*

To *charge* a storage battery, a dynamo must be connected with the lead plates and after it is charged it will, in turn, deliver a constant current. To make the lead plates more active, holes, or grooves, are drilled or cut in them, as shown in Fig. 72. The negative plates are filled with *spongy lead,* and those in the positive plates are filled with *red oxide of lead.*

A storage battery cell is built up of several plates and each positive plate is set between two negative plates. This is to keep the positive plate from warping, or *buckling,* as it is called, when the cell is charged. A *separator* made of thin wood is placed between each positive and negative plate to keep them the right distance apart. A number of the plates, or *groups,* are then assembled into an *element* and set into a jar containing the *electrolyte,* that is, a solution made of pure sulphuric acid and water.

All of the negative plates of a group are connected together and all of the positive plates of a group are connected together and all together they form an *element.* Finally, an element in a jar filled with electrolyte forms a *cell,* see Fig. 72, and two or more cells connected together constitute a *battery.*

How to Use a Storage Battery.—Bear in mind these two things, first: (1) that the number and the size of the lead plates determine the amount of current, or amperes, that the battery can be charged

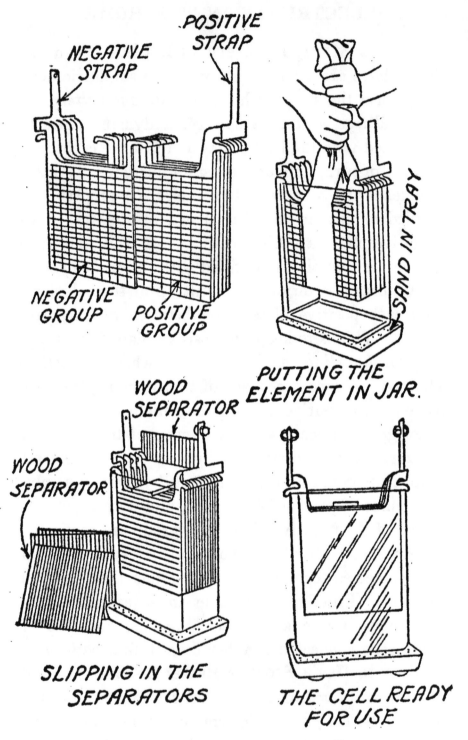

NEGATIVE STRAP

POSITIVE STRAP

NEGATIVE GROUP

POSITIVE GROUP

SAND IN TRAY

PUTTING THE ELEMENT IN JAR.

WOOD SEPARATOR

WOOD SEPARATOR

SLIPPING IN THE SEPARATORS

THE CELL READY FOR USE

FIG. 72.—THE PARTS OF A STORAGE BATTERY

with and will deliver and (2) that each cell has an electromotive force, or voltage, of 2 volts, regardless of the number of plates and the sizes of them. For this reason the voltage is constant and the current varies according to the load.

A storage battery for home lighting and power circuits is made up of 16 cells, and these give 32 volts. But batteries can be had in several different sizes, so that you can store up enough current to light as many lamps at one time or for as long a time without recharging it as you may need.

A battery is rated by the number of *ampere hours* it will give. Thus, a 44 ampere hour battery will deliver 1 ampere for 44 hours or 44 amperes for 1 hour or any *mean,* that is, the equivalent of these figures, according to the load it must take care of.

As an example, a 32-volt, 16-candle power lamp takes about $1\frac{3}{4}$ amperes to light it. Hence you can keep 1 lamp lit on a 44 ampere hour battery for 25 hours, or 5 lamps of the same candle power for 5 hours, without recharging the battery.

The Switchboard and Its Instruments.—The next thing is to connect the dynamo with the storage battery and lights and other apparatus that uses the current. This is done through the *switchboard.*

This is a panel of hard fiber, or better, of slate, and it has on it (1) a *voltmeter,* (2) an *ammeter,* (3) a *rheostat,* (4) an *automatic cutout,* (5) a *double-throw switch* and (6) a pair of *enclosed fuses.*

The connections of the whole installation are shown diagrammatically in Fig. 73.

The voltmeter is an instrument that shows at a glance if the dynamo is generating current and the storage battery is delivering its full voltage. The ammeter tells how much current your lights and other apparatus are using.

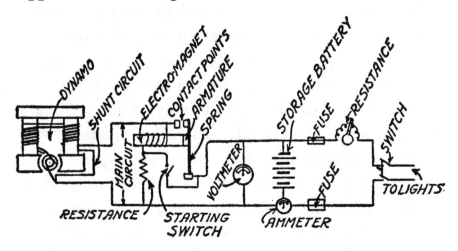

Fig. 73.—Wiring Diagram of a Storage Battery System

The rheostat is a *variable resistance*. By turning a small hand wheel you can cut in or out more or less resistance wire and so regulate the voltage of the dynamo and keep it constant, should the speed of the engine vary too much. The automatic cutout, or *current breaker,* is a switch that closes the circuit which connects the dynamo and the storage battery, when the latter needs recharging, and opens the circuit when the battery is fully charged.

There is also a double-pole, double-throw switch;

when you want to start the engine pull the handle down and it closes the dynamo and storage battery circuit. The current from the latter flows into the former and runs it as a motor. This in turn starts the engine. After the engine is started, you throw the switch up and, when the dynamo is generating current at its full voltage, the automatic cutout closes

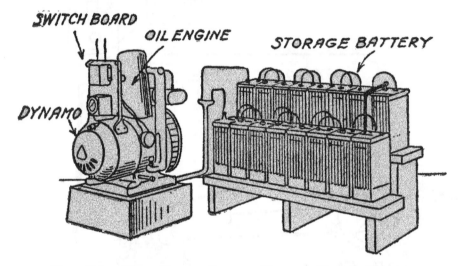

FIG. 74.—THE DELCO-LIGHT DIRECT DRIVE DYNAMO

the circuit and the current begins to charge the storage battery.

The fuses are used to protect the apparatus from surges and overloads. They are made of a lead and tin alloy which melts at a low temperature. Figs. 74 and 75 show two different types of home electric power plants.

Wire for the Transmission Line.—For a 32 volt installation the distance between the power plant and the place where the current is used should not be more than 500 feet, because there is a *drop of voltage* on the line no matter how large the wires forming it may be.

Where a greater distance than 500 feet is to be covered, a 110 volt installation must be used. Bare copper or aluminum wire,[1] or insulated copper wire,

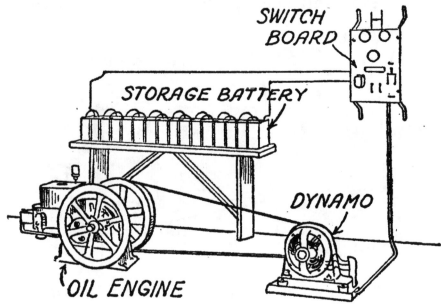

FIG. 75.—THE MORSE FAIRBANKS BELT-DRIVEN DYNAMO

supported on porcelain or glass petticoat insulators,[2] can be used for the transmission lines.

Out and Inside Wiring.—For wires that are to run between buildings, use approved *weatherproof wire;* and for inside wiring use *rubber-covered wire.*

[1] For prices send to the Aluminum Company of America, 120 Broadway, New York City.

[2] For prices write to the Manhattan Electrical Company, Park Place, New York City.

Inside wire must be fastened to the walls either with porcelain knobs, so that they will be kept 1 inch away from the wall, or put in wood or metal molding made for the purpose, or else run between the walls in metal ducts.[3]

The lamps, heating apparatus and motors must be connected across the main line circuit, as shown in Fig. 76, or in *parallel*, as it is called.

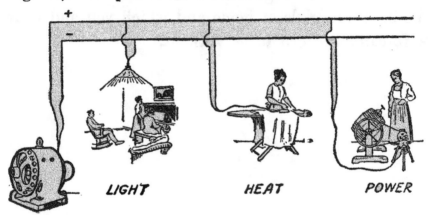

FIG. 76.—LAMPS, HEATING APPARATUS AND MOTORS ARE CONNECTED UP IN PARALLEL

What an Electric Plant Will Do.—Above all it will (1) light your home, barn and grounds and make life worth living; then (2) it will give you heat

[3] Before buying any kind of electrical equipment or doing any kind of wiring, write to the *National Board of Fire Underwriters*, 76 William St., New York City, for a booklet called the ''National Electric Code,'' which will be sent you free of charge. Also write to the *Manhattan Electric Co.*, Park Place, New York City, for a catalogue of materials approved by the above Board for wiring. A very complete description of how to do electric wiring is given in ''The Book of Electricity,'' by the present author and published by D. Appleton and Company, New York.

for curling irons, flatirons, percolators, toasters and other utensils and conveniences; (3) it will give you power to run a tumble churn, coffee mill, cream separator, dishwasher, grindstone, horse clippers, ice-cream freezer, massage vibrator, meat grinder, milking machine, pumps, sewing machine, vacuum cleaner, washing machine, milking machine, etc.; and, finally (4) all of these things make for a life which will keep the boys and girls at home and which you and your wife cannot afford to be without.

CHAPTER XII

USEFUL RULES AND TABLES

TABLE 1

Number and Weight of Pine Shingles to Cover 1 Square of Roof

1 square = 10 × 10 feet or 100 square feet

Number of inches exposed to weather..................	4	4½	5	5½	6
Number of shingles per square of roof....................	900	800	720	655	600
Weight of shingles on 1 square in pounds..........	216	192	173	157	144

(The number of shingles per square is for common gable roofs. For hip roofs add 5 per cent to the above figures. A bundle contains 250 shingles and 1000 four-inch shingles weigh 240 pounds.)

TABLE II

Amount of Water that Can be Raised per Hour by Man, Horse and Wind Power

Power	25 feet	50 feet	100 feet
Man with frequent rests.............	600 gallons	800 gallons	150 gallons
8 foot windmill.....	810 "	400 "	325 "
12 " "	2,400 "	1,320 "	685 "
Horse on treadmill...	7,050 "	3,200 "	1,760 "

TABLE III
Size, Length and Number of Shingle Nails to the Pound

Size	Length and Gauge		Approximate Number to Pound
	Inches	Number	
3d	$1\frac{1}{4}$	13	429
$3\frac{1}{2}$d	$1\frac{3}{8}$	$12\frac{1}{2}$	345
4d	$1\frac{1}{2}$	12	274
5d	$1\frac{3}{4}$	12	235
6d	2	12	204
7d	$2\frac{1}{4}$	11	139
8d	$2\frac{1}{2}$	11	125
9d	$2\frac{3}{4}$	11	114
10d	3	10	83

TABLE IV
Average Weights and Volumes of Fuels

Anthracite coal	1 cubic foot weighs 55 to 65 pounds, 1 ton (2240 pounds) = 34 to 41 cubic feet
Bituminous coal	1 cubic foot weighs 50 to 55 pounds, 1 ton (2240 pounds) = 41 to 45 cubic feet
Charcoal	1 cubic foot weighs 18 to $18\frac{1}{2}$ pounds, 1 ton (2240 pounds) = 120 to 124 cubic feet
Coke	1 cubic foot weighs 28 pounds, 1 ton (2240 pounds) = 80 cubic feet

(One bushel of anthracite coal weighs on an average of 67 pounds; one bushel of bituminous coal 60 pounds; one bushel of charcoal 20 pounds; and one bushel of coke 40 pounds.)

USEFUL RULES AND TABLES

TABLE V

Some Useful Arithmetical Rules

Knowing Diameter to Find Circumference of a Circle.
(1) Multiply the diameter by 3.1416, or (2) divide the diameter by 0.3183.

Knowing Circumference to Find Diameter of a Circle.
(1) Multiply the circumference by 0.3183, or (2) divide the circumference by 3.1416.

Knowing Circumference to Find Radius of a Circle.
(1) Multiply the circumference by 0.15915, or (2) divide the circumference by 6.28318.

To Find the Area of a Circle.
(1) Multiply the square of the radius by 3.1416, or (2) multiply the square of the diameter by .7854, or (3) multiply the square of the circumference by .07958, or (4) multiply the circumference by ¼ of the diameter.

To Find the Area of a Sector of a Circle.
Multiply the length by ½ of the radius.

To Find the Area of the Solid Part of a Ring.
(1) Subtract the area of the inner circle from the area of the outer circle, or (2) multiply the sum of the diameters of the two circles by the difference of the diameters and the product obtained by .7854.

To Find the Area of an Ellipse.
Multiply the product of the two diameters by .7854.

To Find the Area of a Triangle.
Multiply the base by ½ of the altitude.

185

THE AMATEUR MECHANIC

To Find the Area of a Parallelogram.
Multiply the base by the altitude.

To Find the Area of a Trapezoid.
Multiply the altitude by $\frac{1}{2}$ the sum of the parallel sides.

To Find the Area of a Trapezium.
Divide the figure into two triangles, find the area of the triangles, and add them together.

To Find the Surface of a Sphere.
Multiply the diameter by itself, that is, square it, and then multiply this product by 3.1416.

To Find the Volume of a Sphere.
Multiply the diameter by itself twice, that is, cube it, and then multiply this product by 3.1416 and divide the quotient by 6.

To Find the Volume of a Cylinder.
Multiply the diameter of the tank, or other cylinder, by itself, that is, square it; multiply this product by .7854 and, finally, multiply this last product by its height.

INDEX

INDEX

189

INDEX

Formula for finding,
  area of a sector of a
    circle, the, 185
    solid part of a ring,
      185
    trapezium, a, 186
    trapezoid, a, 186
    triangle, a, 185
  circumference of a cir-
    cle, 185
  diameter of a circle,
    185
  heating surface of a
    steam boiler, 130
  height of buildings, 98
  horse power of
    boiler, a, 129
    driving a machine, for,
      91
    electric current, 173
    internal combustion
      engines, 146
    steam engine, a, 130
    water wheel, a, 109
  ohms (resistance), 168
  radius of a circle, 185
    belt needed, a, 80
    gears, 86
    pulley, a, 79
  v o l t s (electromotive
    force), 168
  volume of,
    cylinder, a, 186
    sphere, a, 186
  watts generated, 173

Formula for finding capacity of a water tank,
  54
Frame of a building, 30
Framing square, carpen-
  ter's, 7
Framing table, rafter, 12
Freezing,
  how to prevent water
    pipes from, 60
  mixtures, 158
  point, 65
Friction alloys, anti-, 87
Friction, 87
  how to reduce, 87
  lubricants to reduce, 89
  rolling, 89
  sliding, 87
  what it does, 87
Friction drive (transmis-
  sion), 152
Frost box, how to make a,
  60
Frozen, what to do when a
  water pipe is, 61
Fuel, 135
  hot air engine, for, 135
  reservoir for oil engine,
    143
Fuels, table average, weights
  and volumes of, 184
Fuses, enclosed, 177

Galvanometer, 171
Gas engine, 135

INDEX

INDEX

INDEX

INDEX

INDEX

INDEX

INDEX

Mortar,
  sand for, 34
  stonework, for, 37
Motor car, how to use as a
  power plant, 152

National Board of Fire Underwriters, 146-181
National electric code, 181
Needle nozzle for water
  wheel, 101

Octagon or 8-square, lay-
  ing out an, 9
Ohm, unit of resistance,
  168
Oil engine,
  economy of operation,
  145
  fuel reservoir for, 143
  how to start an, 144
  mixing valve for, 143
  throttling governor for,
  144
Oil engines,
  how to find horse power
  of, 145
  how they work, 142
Oil power, how to use, 151
Overshot water wheel, 100

Packing for stuffing box-
  es, 129
Parallelogram, to find area
  of, 186

Pasteur water filter, 48
Pattern maker's shrinkage
  rule, 5
Pelton water wheel, 101
Penstock for water wheel,
  103
Phosphor-bronze bearing
  metal, 88
Piling for buildings, 26
Pipes from freezing, how to
  prevent water, 60
Pitch of rafters, 12
Pitches, table of common,
  13
Placing concrete, 43
Plain gears, 83
Planimeter, the, 23
Plaster for walls, 34
Plastering for buildings, 26
Plumb, the, 20
Plumb glass, 20
Plumbing,
  a word on, 61
  iron pipe for, 61
  red lead for joints, 61
  table of iron pipe sizes
  for, 62
Pneumatic water supply
  system, 51
Portland cement,
  Atlas, 45
  concrete, for, 45
  made, is how, 41
  stucco, for, 38
  tested, is how, 41

INDEX

INDEX

INDEX

INDEX

INDEX